BIOASSESSMENT OF FRESHWATER ECOSYSTEMS

Using the Reference Condition Approach

The cover shows three streams in Reference Condition (within the ellipse) and one test stream (outside the ellipse), all in the upper Yukon River basin near Whitehorse, Yukon Territory, Canada.

BIOASSESSMENT OF FRESHWATER ECOSYSTEMS
Using the Reference Condition Approach

by

Robert C. Bailey
Department of Biology
The University of Western Ontario, Canada

Richard H. Norris
Cooperative Research Centre for Freshwater Ecology
University of Canberra, Australia

Trefor B. Reynoldson
National Water Research Institute, Environment Canada
Acadia Centre for Estuarine Research
Nova Scotia, Canada

SPRINGER SCIENCE+BUSINESS MEDIA, LLC

Library of Congress Cataloging-in-Publication Data

A C.I.P. Catalogue record for this book is available
from the Library of Congress.

Title: Bioassessment of Freshwater Ecosystems: Using the Reference
Condition Approach
Editor: Robert C. Bailey, Richard H. Norris and Trefor B. Reynoldson

ISBN 978-1-4613-4705-7 ISBN 978-1-4419-8885-0 (eBook)
DOI 10.1007/978-1-4419-8885-0

Printed on acid-free paper.

Contents

Preface

This book is a small but we hope useful contribution to the vigorous field of biological assessment. It deals with the fundamental issue of natural variability, and accepting and using this variability to enhance, rather than detract from, assessments. Although this project has not fully consumed the eight or nine years we have worked on it, we have constantly had "the book" as a backdrop to the other aspects of our respective careers. We celebrate its completion, finally, but are a little wistful that it has ended, since we have so much enjoyed our collaboration over an extended time and space continuum, including a major relocation by one author.

The idea hatched in the wake of our collaboration with Kristin Day, Mike Zarull, and many others on the nearshore Great Lakes environmental assessment (one of the case studies discussed in this book). We thought it would be useful, and frankly pretty straightforward, to produce a how-to guide for the Reference Condition Approach to bioassessment. Work continued rather sporadically, since Bob Bailey was based in London, Ontario, Trefor Reynoldson was in Burlington, Ontario (and then Wolfville, Nova Scotia), and Richard Norris was in Canberra, Australia. We always made a little progress at the North American Benthological Society conference each year, particularly in Kalispell, Montana when we presented "the larval book" in a technical workshop. For our occasional, multi-day work sessions together, we have enjoyed the unbridled hospitality of many friends and colleagues, notably Dave Allen in Ann Arbor, Vince and Cheryl Resh in Berkeley, California, and Jan Ciborowski and Lynda Corkum in Windsor, Ontario.

We decided to bring the proverbial ship into port last year, when Richard had an opportunity to spend his sabbatical at The University of Western Ontario with the help of a Helen Battle Visiting Professorship. Simon Linke was also around and about, re-doing many of the analyses of the case studies to make sure one could do what we were telling people to do. Trefor came for the last few passionate arguments and re-writes, and then we sent the completed manuscript to our scientific editor, Ann Milligan. Ann did a super job determining where we made no sense (we don't fault her for any remaining problems!), and Lindy Paul at Kluwer kept prodding us just enough to get the book into print. We also incorporated many useful comments from Richard Marchant on a near to final draft. On a personal note, we would like to thank Chris Bailey, for putting up with Bob and occasionally the less annoying Richard and Trefor, Ursula Norris, for constant support with a positive outlook that we were doing something useful, and Barbara Reynoldson, who cheerfully provided an environment to help progress and often took wayward co-authors into her and Trefor's home as we completed the book.

Robert Bailey
London Ontario

Richard Norris
Canberra Australia

Trefor Reynoldson
Wolfville Nova Scotia

September 2003

Chapter 1

BIOASSESSMENT OF FRESHWATER ECOSYSTEMS

Fundamental Themes

1. INTRODUCTION

Freshwater ecosystems have a primary role in the biosphere as conduits of water and nutrients from the continents to the sea. They also support unique and complex ecological communities and often define the structure and functioning of the surrounding terrestrial ecosystem. Because of this, and the critical role of fresh water as a human resource, ecologists are often asked to assess or monitor the "health", "status" or "condition" of freshwater ecosystems.

Usually, some of the lakes, streams or wetlands in a region have been exposed to stressors, be they directly (e.g., industrial effluent) or indirectly (e.g., invader species) a result of human activity. An investigation of possibly degraded ecosystems may be part of a general "report card" for a region's streams or lakes, or may be more specifically targeted at sites that are of concern.

There are five stages in such investigations:

1. recognition of a need or mandate to assess freshwater ecosystems in a region;

2. development of a conceptual model to explain how the ecosystems work, including their major potential stressors;
3. design and execution of an assessment study;
4. analysis and interpretation of data from the study; and
5. action by management in response to the interpretation of the study results.

These five components of assessment are usually the responsibility of disparate individuals or groups, often with different or even conflicting perspectives (Cullen 1990; Policansky 1998). The need for an assessment of ecosystems is often first articulated by people who cannot understand or communicate with study designers. The ecologists charged with the task of setting up assessment programs often have great difficulty developing clear study designs to answer straightforward questions about ecosystems (Peters 1993). The ecological significance of the results of such studies is often determined awkwardly and arbitrarily by statisticians or statistical ecologists more concerned and skilled with the analysis than with the interpretation of data. The interpretations are then passed on to those with the mandate to define or alter the course of management actions. They often do not understand any of the other perspectives and, perhaps ironically, do not feel they require input from the other perspectives when making decisions.

Part of the confusion among the various players in assessment and monitoring studies may result from ambiguity in the various approaches used. In this book, we hope to reduce ambiguity and confusion as we add to the large body of work concerned with the assessment of freshwater ecosystems (e.g., Rosenberg and Resh 1993; Karr and Chu 1999; Wright et al. 2000; Downes et al. 2002). We describe a new approach for biological assessment: the Reference Condition Approach (RCA). The RCA measures the variability in biota among sites in "Reference Condition". These reference sites are minimally exposed to human stressors such as effluent discharges or land use changes that often degrade ecosystems. The RCA then uses empirical modeling to explain as much of the variability in the biota of reference sites as possible, based on the environmental characteristics of the sites. The resulting model then predicts the biota that could be expected at a test site, were the site in Reference Condition. In contrast to the reference sites, the test site *is* exposed in some degree to stressors, and it is the effect of the stressors on the biota that is of interest in the assessment.

Although the RCA is applicable to a broad range of ecosystems in both aquatic and terrestrial contexts, our concern in this book is freshwater lakes and streams, and the use of their benthic macroinvertebrates as the biological component of the assessment.

The RCA has a number of key features.

(a) *It defines and quantifies ecosystem health*: The Reference Condition defines empirically (rather than absolutely, theoretically or even metaphysically; cf. Rapport 1989) the condition that we consider to be "healthy" or "acceptable" in an ecosystem. The Reference Condition is established by sampling a number of sites that are minimally exposed to the stressor(s) of interest. The variation among biota in Reference Condition can then be measured (Figure 1-1).

(b) *It explains some variation among healthy ecosystems*: Variation in the Reference Condition of an ecosystem is correlated with, and therefore predictable from, environmental attributes measurable at a wide range of spatial and temporal scales (Figure 1-2).

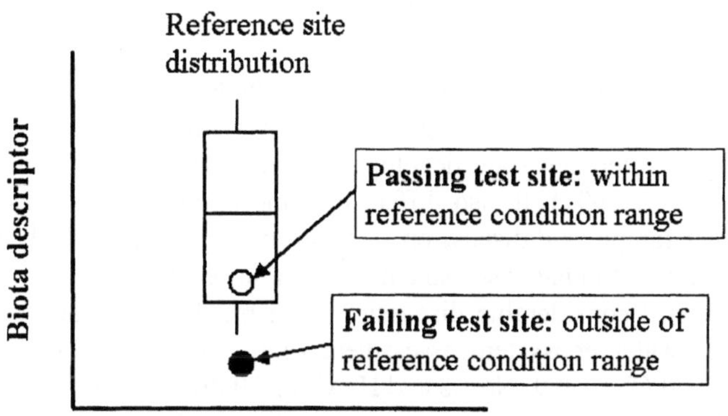

Figure 1-1. Is the biota descriptor at a test site within or outside the range of biota in Reference Condition?

(c) *The deviation of a test site from Reference Condition is a measure of the effect of stressors on the ecosystem*: The final step in an RCA assessment is to check the characteristics of an ecosystem that is exposed to stressors in comparison to the characteristics of a similar ecosystem in Reference Condition. Differences between the biota observed and expected are considered to result from an impact of the stressor(s) (Figures 1-1, 1-2). Ecologically significant sites, worthy of protection, may deviate from Reference Condition in a positive direction (e.g., have greater diversity than expected).

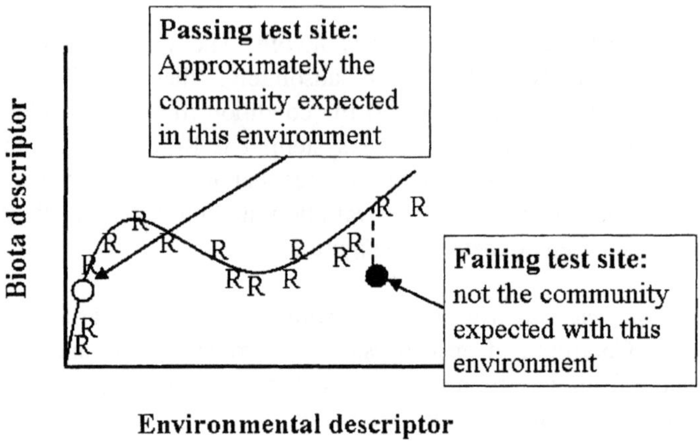

Figure 1-2. Given the environment, is the biota at a test site within or outside the range of communities in Reference Condition?

As we indicate in section 3 of this chapter, the RCA has clear roots in the work of the British RIVPACS group (e.g., Wright et al. 2000), and the comparison to reference state developed by Karr and others in the United States (e.g., Karr 1991). It also differs in some important ways, as we will illustrate by example and discussion — and it diverges significantly from the strong tradition of "impact assessment" in aquatic ecosystems.

There is still considerable discussion on a number of issues within the area of freshwater ecosystem assessment (e.g., Gerritsen 1995 *vs*. Norris 1995; Karr and Chu 1999; Norris and Hawkins 2000). Nominally, this debate concerns details of field and laboratory procedures or data analysis. Actually, it usually reflects deeper divisions in perceptions of and philosophical approaches to the task of bioassessment. Perhaps the combination of assessing human effects on ecosystems, field ecology, hypothesis testing and making the associated statistical analyses, together with a final interpretation leading to management decisions, is really too much for an objective, dispassionate scientific debate to bear. The various players sometimes seem to arrive at their attitudes and approaches in an evolutionary fashion; partly adaptation by natural selection, but with a generous helping of genetic drift and founder effect, along with some chance mutations along the way. To some extent, this has been true of the authors of this book. Indeed, we cannot present ourselves as the objective definers of the true and right path to the evaluation of freshwater ecosystems. In this work we have tried to lay out our approach with sufficient detail and explanation so that those who agree will know how to do what we suggest and those who disagree will know exactly where our downfall is.

2. KEY CONCEPTS

Several concepts are pertinent to both the various components of an ecosystem and the RCA. We present these with little elaboration, and spend the remainder of the book illustrating their application in the context of the assessment and monitoring of freshwater ecosystems.

(a) Freshwater ecosystem: A freshwater ecosystem may have running water (e.g., creeks and rivers) or standing water (e.g., lakes and ponds). It has abiotic physical and chemical attributes as well as biotic communities that reside in it (e.g., benthic macroinvertebrates, fish, meiofauna, macrophytes, algae) over a defined period. The ultimate boundary of a *site* in a freshwater ecosystem is the boundary of its catchment area. In principle, the structure and function of a freshwater ecosystem at a given point can be influenced by conditions and processes anywhere within (or sometimes even beyond) its catchment area (see Hynes 1975). For example, even the effects of large-scale air pollution will be largely channeled to a site through its catchment. Adequate description of the habitat or environment of the biota in the ecosystem may cover a range of spatial and temporal scales. At one extreme is a single point and moment of sampling (e.g., mean flow rate over one minute at a point in a stream). At the other extreme is a process or feature characterized for the entire catchment area over a long period (e.g., mean total precipitation over several years in the catchment area).

(b) Sites in a freshwater ecosystem: There has been a growing realization that the spatial scale, or more precisely, the grain of hypotheses we wish to test about ecosystems exposed to stressors is the site itself, rather than points within a site. We used to measure differences between a set of "control" (what we would call "reference") and a set of "impact" (what we call "test") sites. Now, typically, our task is to evaluate one or more "test" sites and determine whether or not they are within the range of variability among "reference" sites. We use reference sites where the scale of replication is the site itself, not sub-samples within it. So habitat and biota descriptors, while they may be measured at various spatial and temporal scales, all ultimately refine our description of a reference or test site. As we discuss in Chapter 3, there are no absolute constraints on the

spatial proximity of an array of sample sites. Sites within the same lake or stream, or indeed in other freshwater systems in the region, may be from meters to kilometers away from each other, depending on the spatial extent of the study and the resources available for evaluation within that area. There is no particular spatial scale at which sites within a lake or stream, or a set of lakes or streams, become ecologically independent. Adjacent points in a reach, reaches in a stream, streams in a catchment, or catchments in an ecoregion may, to a greater or lesser extent, have similar biological communities because of their "ecological proximity". This contrasts with the view sometimes held that spatial autocorrelation almost always constrains study design and statistical analysis in observational ecology (e.g. Nash et al. 1999).

(c) *The Reference Condition*: Sites that are thought to have been minimally exposed to the stressor(s) of interest are in a state that we refer to as Reference Condition. Specific, operational criteria must be set that define reference sites by their lack of exposure to the stressor(s) of interest (see Chapter 3). We reject subjective definitions of the ecosystem health of a test site that include qualities such as diversity of biota, interactions among species and other, often unmeasurable, attributes of communities within the ecosystem (cf. Rapport 1989). Reference sites encompass the range of biota variability among sites in the region being assessed that are minimally exposed to the stressor(s).

(d) *Ecosystem assessment*: We usually evaluate "test" sites that we suspect may have been either detrimentally affected by exposure to a stressor (e.g., effluent from a pulp and paper mill) or improved by some amelioration of stressor(s) (e.g. enhanced wastewater treatment). We consider such sites to be "exposed" to the stressor, not "impacted" or "degraded". In other words, we *hypothesize* that a *change* has occurred in the ecosystem in response to exposure to the stressor(s). If we detect a difference between the observed and predicted (Reference Condition) biota at the test site, we require further study of the site, or at least examination of additional data (see Chapter 6), to determine whether or not this change has been *caused* by exposure to the stressor. Further evidence, from examining the specific nature of the change (e.g. species known to be sensitive to the stressor(s) have changed in abundance), or deviation of the site's environment from Reference Condition, may be consistent with the hypothesis that exposure to the stressor has caused the change, in which case some specific management actions may be warranted. Assessments can consist of one-time surveys of reference and test sites, or repeated surveys of such sites (surveillance), or surveillance with a goal of evaluating change through time relative to some established criterion (monitoring) (modified from Helawell 1986). Occasionally, assessment is used to identify sites of exceptional importance that are worthy of protection. In this case, we would still be testing the hypothesis that a site deviates from Reference Condition, but the test site might have, say, exceptionally diverse biota, or rare species.

(e) *Benthic macroinvertebrates in bioassessment*: The organisms most frequently used for biological assessment of freshwater ecosystems are benthic macroinvertebrates (Hawkes 1979; Wiederholm 1980; Abel 1989). Benthic macroinvertebrates are used in over 90% of United States (US) assessment programs (Diamond et al. 1996). Although most assessment programs were originally developed using fish as the biotic component, only about half of the state water quality programs include a fish component, usually in conjunction with the benthic macroinvertebrate community. Arguments for and against using macroinvertebrates in bioassessments have been clearly articulated elsewhere (e.g., Rosenberg and Resh 1993). Here, we highlight the qualities of the macroinvertebrate community that we consider to be the most important. First, the benthic invertebrate taxa present at a site can be identified relatively easily, and they are present in almost all freshwater ecosystems. Also, they are easier to collect than biota that are perhaps more directly relevant to human concerns (such as fish) or that are more diverse and abundant (such as micro-organisms). Second, many members of the macroinvertebrate community are known to change in abundance with changes in common stressors of interest (such as organic pollution, metals), and therefore the provide a useful tool in judging the condition of the ecosystem. Third, benthic macroinvertebrates are relatively sedentary and provide a good spatial signal of what has occurred at the site unlike fish, for example, which may have been exposed to a stressor tens or even hundreds of kilometers away from where they are collected.

(f) *Predictive modeling of the Reference Condition*: One of the most important features of the RCA is the modeling or "explanation" of biotic variation across the reference sites, using environmental descriptors. The biota at an exposed site are not just compared to the mean of those at all reference sites (with their associated variability). Instead they are compared to the biota expected, given the empirically determined predictive model that relates reference biota to the environments in which they are found. This can often result in a more sensitive comparison between a test site and Reference Condition than the classical, "upstream *vs.* downstream" comparison (Green 1979), with its potential for confounded covariation of the communities and the stressor(s). In Chapter 4, we will illustrate several approaches to this modeling, which "sensitizes" the monitoring assessment in direct proportion to the amount of variation among reference communities that can be explained by environmental variation.

(g) *Determining if a test site differs from Reference Condition*: In recent ecological studies, particularly in assessment, the issue of ecological *vs.* statistical significance has been widely debated. In the context of monitoring and assessment, the debate is largely concerned with the establishment of *effect size*, that is, the deviation from Reference Condition that we wish to detect, and the relative importance of Type I (mistakenly failing a site) and Type II (mistakenly passing a site) errors (Keough and Mapstone 1997; Quinn and Keough 2002). When considering this issue in detail (Chapter 5) we emphasize careful consideration of the management implications of passing or failing a site. The best consequence of passing a test site is not the conclusion that the site is "OK" and can be ignored in future, but rather the continued monitoring of it. The consequence of failing a test site, rather than shutting down any and all potential sources of stressors, may be further detailed study to find out why the site is not in Reference Condition, thereby providing direction for the best course of management action (Chapter 6).

3. DEVELOPMENT OF THE REFERENCE CONDITION APPROACH

The bioassessment of freshwater ecosystems has evolved through a variety of approaches (Figure 1-3). Here we briefly outline the chronology of freshwater bioassessment from its earliest beginnings to the present day, and highlight the major trends and significant developments over that period.

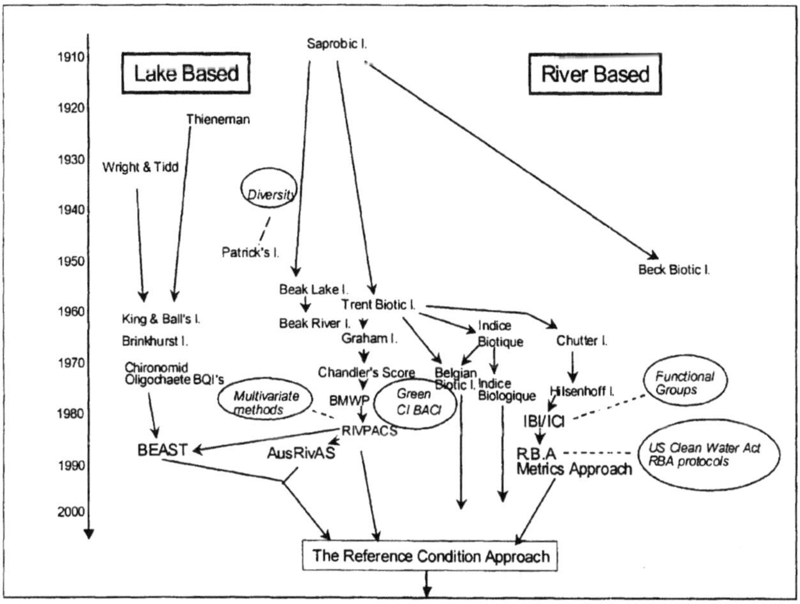

Figure 1-3. The evolution of biological assessment of freshwater ecosystems in the 20th century. (Saprobic System: Kolkwitz and Marsson 1909; Thieneman Lake Classification: Thienemann 1925; Wright and Tidd: Wright 1955; Patrick I: Patrick, 1949; Beak Indices: Beak, T.W. 1965.; King and Ball I.: King and Ball 1964; Brinkhurst I: Brinkhurst 1966; Oligochaete and Chironomid BQIs: Milbrink 1973, Wiederholm 1980, Howmiller and Scott 1977; Beck Biotic I: Terrell and Perfetti 1996; Trent Biotic Index: Woodiwiss, 1964; Chandlers Score: Cook 1976; BMWP: ISO 1979: RIVPACS: Wright et al. 1984 ; BEAST: Reynoldson et al. 1995; AUSRIVAS: Davies 2000; Belgian Biotic I.: De Pauw et al 1979; Indice Biotique: Tuffery and Verneaux 1968; Indice Biologique: Verneaux et al 1982; Chutter I: Chutter 1972; Hilsenhoff I. : Hilsenhoff 1977; IBI: Karr 1981; RBA and Multimetrics: Plafkin et al. 1989).

3.1 The Roots of Bioassessment: 1900–1950

Surface water quality assessment based on biological indicators of pollution began in Europe in the early twentieth century. Burgeoning human populations and industrial development began to severely affect aquatic ecosystems, but also produced a critical mass of well-trained scientists.

The concept of biological indicators of aquatic conditions originated with the work of Kolkwitz and Marsson (1909) in Germany, and their development of the Saprobic System for rivers. This method is mainly based on the presence of microorganisms of the plankton and periphyton community in response to sewage contamination, and the resulting decrease in oxygen. The Saprobic System has been extensively revised by European scientists and is still the basis of some systems currently in use (e.g., Bauernfeind and Moog 2000). It has been argued by some that the Saprobic System never found much popularity in North America (Metcalfe 1989; Cairns and Pratt 1993). However, we suggest the system is really the basis of biotic index approaches, such as number of Ephemeroptera, Plecoptera, and Trichoptera (EPT) taxa and Hilsenhoff's Biotic Index (Hilsenhoff 1988), still in use today as part of the various integrated metrics. These indices have usually been developed empirically as a method for assessment of organic pollution (Norris and Georges 1993).

Many reviews of bioassessment literature leave out much of the early work other than that of Kolkwitz and Marsson (1909), and this sometimes leads to "reinventing of the wheel". By way of example, in 1948, Reynoldson (father of Trefor Reynoldson) published a synthesis of work on enchytraeids in sewage filter beds (Reynoldson 1948). The work was a combination of community-based studies and laboratory experimentation, examining changes in reproduction in the species and linking this to environmental conditions. The method, 40 years later, was formalized as the sediment quality triad (Long and Chapman 1985), which was seen by many as a major step forward in assessing sediment contamination. Another feature of the pre-1950s period was attempts to classify freshwater ecosystems based on the invertebrate communities present, resulting in lake classification (Thienemann 1921, 1925) and river zonation (Forbes and Richardson 1913) systems. The best summary of the approaches used during this period is Hynes's (1960) classic volume, *The Biology of Polluted Waters*.

3.2 Environmental Awareness and Indices: 1950–1970

Two notable events during the period 1950–1970 were i) the rise of environmental awareness and the establishment in many jurisdictions of agencies with environmental mandates, and ii) the development of various biotic indices and scoring systems, including diversity indices.

The development of environment agencies generated a market for formal approaches to bioassessment. This market was part of the impetus for the development of the Trent Biotic Index (Woodiwiss 1964), which was one of the earliest forms of biotic index. It was restricted to six groups of key organisms, and assigned scores to individual taxa based on their responses to certain pollutants. The index was the precursor of many that followed, including the widely used Hilsenhoff index in North America, the BMWP (Biological Monitoring Working Party) score system used in the UK (see ISO 1979; Wright 1995), the Index Biotique (France, De Pauw and Van Hooren 1983), the Belgian Biotic Index (Tuffrey and Verneaux 1968), and so on. The main trend in the development of these approaches was an increase in the numbers of taxa to which scores were assigned, thus making them based more on the entire benthic invertebrate community.

The previous approaches were really extensions of the indicator species approach to assessment. In the late 1940s, Ruth Patrick developed community-based assessments based on her studies of diatom assemblages (Patrick 1950). Patrick's work was influenced by the later-published ideas of MacArthur and Wilson (1967), whose community concepts were based on a dynamic equilibrium of immigration and extinction of species in a community. Applied ecologists turned towards equilibrium theory and diversity measures to describe the essential nature of a community in evaluating conditions in aquatic ecosystems. Initially, diversity indices were attractive and were adopted rapidly (Wilhm and Dorris 1968), particularly by agencies which were enthralled by the idea of reducing the complexity of the biota to a single number. However, the relationship between diversity and stability and its extension, environmental quality, was frequently questioned (e.g., Hurlbert 1971). Diversity indices were dealt a body blow by Poole (1974) who stated that diversity indices were "answers to questions which have not yet been found", and Green (1979), who declared that

> the strongest argument against the use of diversity indices as a derived criterion or predictor variable in environmental studies is that other statistical methods retain more of the information in the biological data while reducing them to a more ecologically meaningful form.

In fact, diversity indices are still used in many assessment studies and were precursors of some of the multivariate methods currently in use that capture the information in biological data in the meaningful way described by Green.

3.3 Statistical Rigor and Rapid Assessment: 1970–present

There has been a vast proliferation of bioassessment literature and research activity during 1970–2002. Important events through this period have included the detailed development of study designs and parametric statistical analysis; the development and, to some degree, standardization of rapid sampling protocols, which include the formalized integration of many indices in the multimetric approach (see Plafkin et al. 1989; Karr and Chu 1999); and the development of multivariate methods and predictive models, now largely available because of the enhanced computing power of personal computers.

The design of bioassessments for aquatic (freshwater *and* marine) ecosystems has evolved from Green's (1979) seminal work on comparison of control and impact sites, before and after start-up of a stressor (the BACI design), and Stewart-Oaten's et al. (1986) repeated sampling of control and impact sites, before and after start-up. Recent developments include relatively complex ANOVA designs (Keough and Mapstone 1997; Underwood 1997). Unfortunately, resource constraints (money, time, political will) force most bioassessments to use the crudest of Green's (1979) designs, so all the refinements of the BACI design developed over the last 25 years have been necessarily ignored by many working "in the trenches". Some of the niceties of ANOVA designs need to be restrained to get satisfactorily precise answers to questions about specific ecosystems.

In North America, there was a strong emphasis in the 1970s and 1980s on the use of parametric statistical analysis (e.g., Green 1979; Allan 1984; Hurlbert 1984). Green's (1979) volume covers sampling design and data analysis. It is still highly relevant and provides excellent advice on good study design and clear expression of study objectives. As noted above, this good advice caused perhaps excessive constraints on study designs, particularly following Hurlbert's (1984) comments on confounded designs of upstream versus downstream assessment studies. A succession of more complicated study designs, many summarized in Downes et al. (2002), tried to address the "pseudoreplication" issue in ecological assessment. But the sophistication and consequent cost of such designs quickly outpaced the perceived or real benefits in practical terms in a timely fashion.

The formalization of rapid bioassessment protocols for running waters in the US (Plafkin et al. 1989) provided much of the impetus for the proliferation of such programs in the late 20th century. Since the initial publication of the protocols, the US Environmental Protection Agency (US EPA) has produced numerous technical documents providing specific technical guidance (US Environmental Protection Agency 1990, 1991a, 1991b, 1991c). Additionally, the US EPA has sponsored both regional and national workshops and symposia to facilitate state efforts to develop biological assessment programs (e.g. US EPA 1991d). The current US EPA policy requires that all states incorporate biological criteria into their water quality standards, and is encouraging the use of biological criteria throughout water resource management activities required under the Clean Water Act. This strategy has been so successful that 47 states and the District of Columbia and Puerto Rico now have some type of biological monitoring in their water resource programs (Resh and Jackson 1993; Diamond et al. 1996), using benthic macroinvertebrates (47 states), fish (25 states) and periphyton (3 states).

Compared to protocols and procedures for running waters, technical guidance for the bioassessment of lakes has made more modest progress in the United States (i.e., US EPA 1998), but still encourages a variation of the RCA. It is recommends that communities found in lakes that are exposed to stressors of some kind should be compared to those in a lake or lakes in Reference Condition.

Green (1971, 1974) used multi-group discriminant analysis for quantifying the multivariate niche of species, and this work was a precursor to the widely used Canonical Correspondence Analysis in water quality assessment summarized by ter Braak and Verdonschot (1995). Green's discriminant analysis approach also provided an analytical platform for the development of predictive models such as RIVPACS (River Invertebrate Prediction and Classification Scheme; Wright 1995).

Multivariate predictive models were first developed for, and applied to, national biological monitoring programs took place in the United Kingdom (UK) beginning with a project in 1977 at the River Laboratory of the Institute for Freshwater Ecology (IFE). The project had two objectives (Wright et al. 1984): to classify unpolluted running-water sites based on benthic macroinvertebrates, and to determine whether the biota expected at unstressed sites could be predicted from physical and chemical features of the site. This of course is a statement of the most important attributes of the RCA (variability and predictability in Reference Condition), and thus the UK approach is the direct predecessor of the RCA as we define and implement it. The approach used in the IFE project, linking the structure of the biota to the structure of the environment, followed methods developed by Field et al. (1982). It is being continued in the marine environment (e.g., Warwick et al. 1990; Clarke and Ainsworth 1993).

The classification of reference sites was seen as important by those involved in river conservation as the first stage of a new approach for the detection and assessment of environmental impact. The project originally classified 268 unpolluted sites sampled on 41 rivers. As of 1995 it included approximately 700 reference sites (Wright 1995). The RIVPACS approach was supported by a Royal Commission on Environmental Pollution in 1992, which recommended that regulatory authorities should endeavor to develop a general classification scheme for UK waters based fully on their biological condition. The predictive models were in their third generation (RIVPACS III) by 1995, and were the basis of a national stream assessment in a second, quinquennial, national river quality survey (Wright et al. 1998).

In Canada, there is no ongoing national program using benthic invertebrates to monitor water quality at a regional scale. The RCA has been successfully implemented in Canadian streams of the Fraser River basin, British Columbia, and the near shore, fine sediments of the Great Lakes (see Chapter 2). Both of these studies are used as examples throughout this book. The programs were developed to satisfy separate mandates for bioassessment in these two, large, ecologically distinct areas of Canada. Recently the RCA has been tested on a regional scale in streams (Hawkins et al. 2000) and in wetlands (King and Richardson 2002) in the US.

In continental Europe there have been a few applications of the RCA on a limited, regional scale — in Spain (Rodriguez and Wright 1991), Portugal (Feio et al. 2002) and Sweden (Johnson and Wiederholm 1989; Johnson and Goedkoop 2000).

In Australia, biological assessment of freshwater ecosystems has developed markedly during the last decade. Much of this activity has resulted from concern about ecological values that has led to an emphasis on aquatic biota for assessing conditions directly. A statement on the environment at the end of 1992 by then Prime Minister Paul Keating led to the provision of funding enabling the establishment of the National River Health Program. The mandate of the program was to provide a means of assessing the ecological condition of Australia's river systems. The program was set up to, among other things, assess and monitor the health of Australian rivers.

The National River Health Program adopted a modified version of the RIVPACS program from the United Kingdom. This required a standard, working platform (called the Australian River Assessment System, AUSRIVAS) that would allow water managers to compare the condition of streams and rivers both within and among all Australian states and territories. In 1994–1995 almost 2000 sites were sampled throughout Australia, in spring and autumn, so that the variability of sites in Reference Condition could be assessed, and so that models could be built to predict communities in Reference Condition, from environmental variables. The AUSRIVAS computer program and these models are available on the web (http://ausrivas.canberra.edu.au). The study for the Australian Capital Territory, which formed a small component of the National River Health Program, is used in this book to illustrate the RCA. In 1997, Australia embarked on the First National Assessment of River Health, in which about 6000 sites are sampled in spring and autumn each year. This is probably the first continent-wide assessment of river condition.

The apparent, initial success of the National River Health Program in Australia has led to a trial study on the Brantas River in East Java, and the RCA may be adopted for national water quality assessment in Indonesia (Hart et al. 2001; Sudaryanti et al. 2001).

4. POINTS TO REMEMBER

— Ecological assessments provide essential information for management and protection of freshwater ecosystems.
— The RCA enables an empirical definition of ecosystem health.
— Healthy ecosystems vary in their biota and environmental features at a wide range of spatial and temporal scales.
— The covariation of the biota and environmental features of healthy ecosystems enables the construction of predictive models
— Predictive models can measure an ecosystems deviation from Reference Condition, which in turn is a measure of the effect of stressors on the ecosystem.
— To identify the cause(s) of an ecosystem's deviation from Reference Condition, we require additional analysis.
— The RCA differs from traditional impact assessment by employing designs and analysis that avoid many of the statistical and resource constraints of more traditional ANOVA designs.
— Biological assessment has evolved over the last century, and the RCA has emerged as broadly applicable in many contexts.

Chapter 2

INTRODUCTION TO THE CASE STUDIES

Examples of RCA studies

We use three case studies to demonstrate the use of the Reference Condition Approach (RCA). The studies represent three markedly different environments: the lentic, near shore environment of the North American Great Lakes; tributaries of the Fraser River, a large, west-coast basin in Canada; and the upper Murrumbidgee River basin in the Australian Capital Territory.

The Great Lakes and Fraser River studies had very large spatial extents (tens of thousands of square kilometers) relative to the upper Murrumbidgee River (hundreds of square kilometers). However, all three studies cover areas with clear natural environmental gradients and a wide variety of habitat types. All three studies were designed following strong mandates from government and communities, and with collaborative input from the parties involved, and each of the case studies had objectives leading to clear practical outcomes, where the goals and how to meet them were determined jointly. Each study also had well defined criteria for site selection and standardized methods for sampling and analysis.

1. SEDIMENT ASSESSMENT OF THE NEAR SHORE ENVIRONMENT OF NORTH AMERICAN GREAT LAKES

The Great Lakes in North America is the largest lake system on earth, containing roughly 18% of the world's fresh water (Figure 2-1). The magnitude of the Great Lakes water system is difficult to appreciate. The lakes contain about 23,000 km^3 (5500 cu. Mi.) of water, and cover an area of 244,000 km^2 (94,000 sq. Mi.) in total. Only the polar ice caps contain more fresh water.

Figure 2-1. North American Great Lakes. The area photographed is approximately 1000km (wide) x 800km (high). Photo courtesy of NASA Visible Earth project.

Because of the large size of the drainage basin, physical characteristics such as climate, soils and topography vary substantially across it. To the north, the climate is cold and the terrain is dominated by the granite bedrock of the Canadian Shield (Precambrian rocks under a generally thin layer of acidic soils). Conifers dominate the northern boreal forests. In the southern areas of the basin, the climate is much warmer, and the soils are deeper with layers or mixtures of clays, silts, sands, gravels and boulders deposited as glacial drift or as glacial lake and river sediments. Soil is relatively fertile in many areas of the southern basin, thus the original deciduous forests have given way to agriculture and sprawling urban development.

Although part of a single system, each lake has its own unique properties. In volume, Lake Superior is the largest, as well as being the deepest and coldest of the five, and has a correspondingly long retention time of 191 years. It could comfortably contain all the other Great Lakes. Most of the Lake Superior basin is forested, with little agriculture because of a cool climate and poor soils. The forests and sparse population result in relatively few pollutants entering Lake Superior, except through airborne transport.

Lake Michigan, the second largest lake, is the only Great Lake entirely within the United States. The northern part of Lake Michigan is in the relatively cold and undeveloped upper Great Lakes region, which is sparsely populated except for the Fox River Valley that drains into Green Bay. This bay has one of the most productive Great Lakes fisheries but receives the wastes from the world's largest concentration of pulp and paper mills. The more temperate southern basin of Lake Michigan is in one of the most urbanized areas in the Great Lakes system. It contains the Milwaukee and Chicago metropolitan areas. This region is home to about 8 million people, or about one-fifth of the total population of the Great Lakes basin.

Lake Huron, which includes Georgian Bay, is the third largest of the lakes by volume. The sandy beaches of Huron and the rocky shores of Georgian Bay make it a prime recreational area. The Saginaw River basin is intensively farmed and contains the Flint and Saginaw-Bay City metropolitan areas. Saginaw Bay, like Green Bay, contains a very productive fishery.

Lake Erie is the smallest of the lakes in volume and is exposed to the greatest impacts from urbanization and agriculture. Because of the fertile soils surrounding the lake, the area is intensively farmed. The lake receives runoff from the agricultural area of southwestern Ontario and parts of Ohio, Indiana and Michigan.

Seventeen metropolitan areas with populations over 50,000 are located within the Lake Erie basin. Although the area of the lake is about 26,000 km^2 (10,000 sq. Mi.), the average depth is only about 19 m (62 ft). It is the shallowest of the five lakes and therefore warms rapidly in the spring and summer, and frequently freezes over in winter. It also has the shortest retention time of the lakes, 2.6 years. The western basin, comprising about one-fifth of the lake, is very shallow with an average depth of 7.4 m (24 ft) and a maximum depth of 19 m (62 ft).

Lake Ontario, although slightly smaller in area than Lake Erie, is much deeper than its upstream neighbor, and has an average depth of 86 m (283 ft.) and a retention time of about 6 years. Major urban industrial centers, such as Hamilton and Toronto, are located on its shore. The US shore is less urbanized and is not intensively farmed except in a narrow band along the lake.

In spite of their large size, the Great Lakes are sensitive to the effects of a wide range of pollutants. The sources of pollution include the runoff of soils and farm chemicals from agricultural lands, the waste from cities, discharges from industrial areas and leachate from disposal sites. The large surface area of the lakes also makes them vulnerable to direct atmospheric pollutants that fall with rain or snow and as dust on the lake surface. Outflows from the Great Lakes are relatively small (less than 1% per year) in comparison with the total volume of water. Pollutants that enter the lakes, whether by direct discharge along the shores, through tributaries, from land use or from the atmosphere, are retained in the system and become more concentrated with time. Also, pollutants remain in the system because of resuspension (mixing back into the water) of sediment and cycling through biological food chains.

In 1991, a project to evaluate the state of near shore, primarily fine-sediment areas of the North American Great Lakes was launched by Environment Canada. For three years following the planning and design of the study, sampling of benthic macroinvertebrate communities and their environments was carried out at over 250 sites in the Great Lakes. About 95% of these sites satisfied the criterion of being only indirectly exposed to potential stressors (such as industrial waste from outfalls) and other habitat criteria for Reference Condition. Several sites directly exposed to stressor(s) were treated as test sites in this execution of an RCA study.

2. FRASER RIVER (BRITISH COLUMBIA) BIOMONITORING PROGRAM

The Fraser River is British Columbia's largest river, with a catchment area of approximately 230,000 km^2, or about 25% of British Columbia, Canada's westernmost province (Figure 2-2). The Fraser itself has a mean annual discharge of 3620 $m^3\ s^{-1}$, which makes it the sixth largest river in North America (Dynesius and Nilsson 1994).

Approximately 1.7 million people live in the Fraser catchment, (two-thirds of British Columbia's population). The river's headwaters originate in three major mountain ranges: the Rocky, the Columbia and the Coast Mountains. The geological, climatic and landform diversity in the catchment is so great that it includes 11 of the 14 biogeoclimatic zones identified in British Columbia. Although the Fraser is one of the last unregulated large rivers of North America, urban and industrial pressures pose a severe threat to the health of this ecosystem. This threat was recognized by the establishment in 1991 of the Fraser River Action Plan (FRAP) by Environment Canada (Environment Canada 1995).

From its source in the Rocky Mountains, in Mount Robson Provincial Park near the Alberta border, the river flows 1,370 km (850 mi) before emptying, through a delta, into the Strait of Georgia, near Vancouver. The river initially flows northwest through a section of a deep, narrow valley called the Rocky Mountain Trench. It then turns south near the city of Prince George, where it is joined by its major western tributary, the Nechako River. In its central section, the volume of the river increases, and below Quesnel its banks gradually take on a canyon-like aspect. Important tributaries in this section include the West Road and Chilcotin Rivers, from the west, and the Thompson River, from the east. From Lytton to Yale the river flows through a canyon of great scenic beauty. At the canyon's southern end the Fraser River passes between the Cascade Range to the east and the Coast Mountains to the west. A little below Yale, at Hope, the river turns sharply west, and the fertile lower Fraser Valley begins. The Fraser River empties into the Strait of Georgia through three main channels.

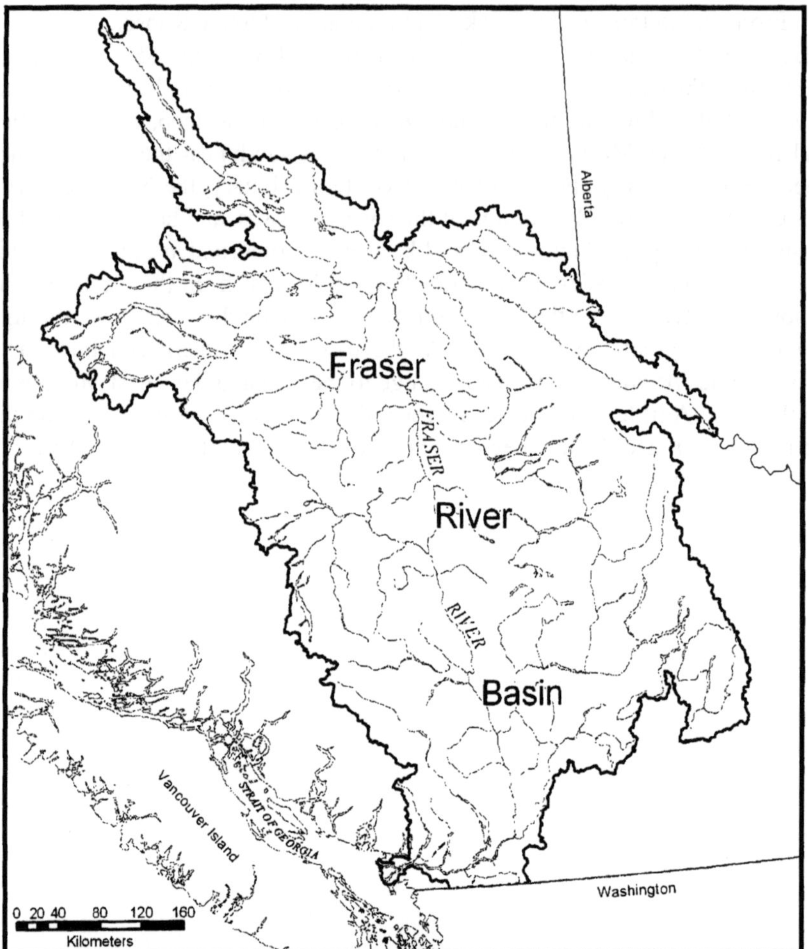

Produced by. Environment Canada, Pacific & Yukon Region, Application & Services. August 7, 2003. Projection: B.C Albers Equal Area Conic NAD 83
Sources Natural Resources Canada, Geomatics Canada National Atlas Digital Data 1:2,000,000, B.C Ministry of Sustainable Resource Management. British Columbia Watershed Atlas 1:50,000
This map is compiled from various sources and is not warranted as to its accuracy, sufficiency, or currency

Figure 2-2. Fraser River (British Columbia) basin.

Much of the river basin is heavily wooded, and forest-products industries, both logging and pulp and paper production, dominate the economy of the settlements along the river. The lower Fraser Valley, including the delta, has highly productive farms. Several species of salmon spawn in the Fraser, and salmon fisheries are located near the river's mouth. While the river has great hydroelectric potential, it remains undeveloped for fear of detrimental effects on the migratory habits of the salmon.

In 1993, an assessment of the state of the Fraser River Basin was undertaken as part of the Fraser River Action Plan (FRAP). This initiative was supported by, and had the participation of, provincial and federal scientists and policy makers. Tributaries of the Fraser River as small as first order (scale 1:50,000) and as large as the main stem of the Fraser River itself were sampled over three years. During the sampling period, over 90% of the streams sampled were considered to be in Reference Condition, while about 10% of the sites were exposed to some sort of stressor (e.g., pulp and paper mill effluent), and were thus considered test sites for the RCA.

3. SPRING AND AUTUMN ASSESSMENT OF STREAMS IN THE AUSTRALIAN CAPITAL TERRITORY

The upper Murrumbidgee River catchment covers an area of 13,000 km^2 in inland eastern Australia. All rivers and streams in the area ultimately flow into the Murrumbidgee River as it travels north and then west before entering Australia's major inland river, the Murray. At the upstream end of the study area is Tantangara Reservoir, part of the Snowy Mountains Hydroelectric Scheme, 40 km from the river's source. The downstream end of the study area is marked by an irrigation-water storage, Lake Burrinjuck, 320 km from the source. Some of the major tributaries are also dammed. For example, in the western part of the catchment the Cotter River has three water supply storages: Corin Dam, 20 km from the source, Bendora Dam, 40 km from the source and the Cotter Dam, 70 km from the source. In the south-east, the Googong Reservoir water supply storage is located 80 km from the source of the Queanbeyan River, another tributary.

Land use includes sheep and cattle grazing, forestry, urban settlement and national parks or reserves managed for conservation or water supply. Most of the national parks are situated in the western part of the catchment, and grazing occurs predominantly in the eastern, northern, and southern parts of the catchment. Canberra, the Australian capital, a city of about 300,000 people, lies in the northeastern part of the catchment (Figure 2-3). The only other urban centers of any size are Queanbeyan, near the junction of the Queanbeyan and Molonglo River tributaries in the east, Cooma, on the upper Murrumbidgee River, and Yass, on the Yass River.

The climate is temperate and characterized by mild winters (average daily maximum temperature = 11°C in July) and warm summers (average daily maximum temperature = 27°C in January). Rainfall is relatively uniform across the upper Murrumbidgee River catchment, with a long-term median ranging from 690 mm in the southeast to 610 mm at Yass in the north, and snowfall is common on the western ranges above 1200 m. However, temperature and rainfall are variable over time and severe droughts and floods occur periodically.

An allocation of $10 million of Australian Government funds in 1992 made possible the National River Health Program (NRHP) in 1993. The main objectives of the NRHP were: to develop and enhance techniques to monitor and assess rivers; to establish a national approach to monitoring and assessing rivers; and to conduct the First National Assessment of River Health (FNARH). The success of this program has led to the development of the Australian River Assessment System (AUSRIVAS) which is based on the British RIVPACS methods (Wright et al. 1984, Moss et al. 1987, Wright 1995). AUSRIVAS consists of standardized field and laboratory methods and taxonomic materials, including procedures for quality control and quality assurance and training and accreditation, for each Australian state and territory and a centralized Internet-based software package for analysis of the data (http://ausrivas.canberra.edu.au). AUSRIVAS was formally adopted as an indicator for national State of the Environment reporting (Ball et al. 2001) and for the National Land and Water Resources Audit (Norris et al. 2001). It has been incorporated into the Australian National Water Quality Guidelines (ANZECC & ARMCANZ 2000) and is being progressively incorporated into state licensing and monitoring guidelines.

Stream sampling was carried out at about 2000 reference sites, including almost 200 sites the Australian Capital Territory (ACT). ACT streams from first to fifth order were sampled in riffle and pool habitats in spring and autumn over three years. Most (over 85%) of the streams fulfilled the criteria defined for Reference Condition (see Chapter 3). Others were exposed to various urban and industrial stressor(s), and thus were considered test sites for the RCA. In this book, we present results from separate spring and autumn samplings of only the riffle habitat, to illustrate and make a seasonal comparison of RCA performance in bioassessment.

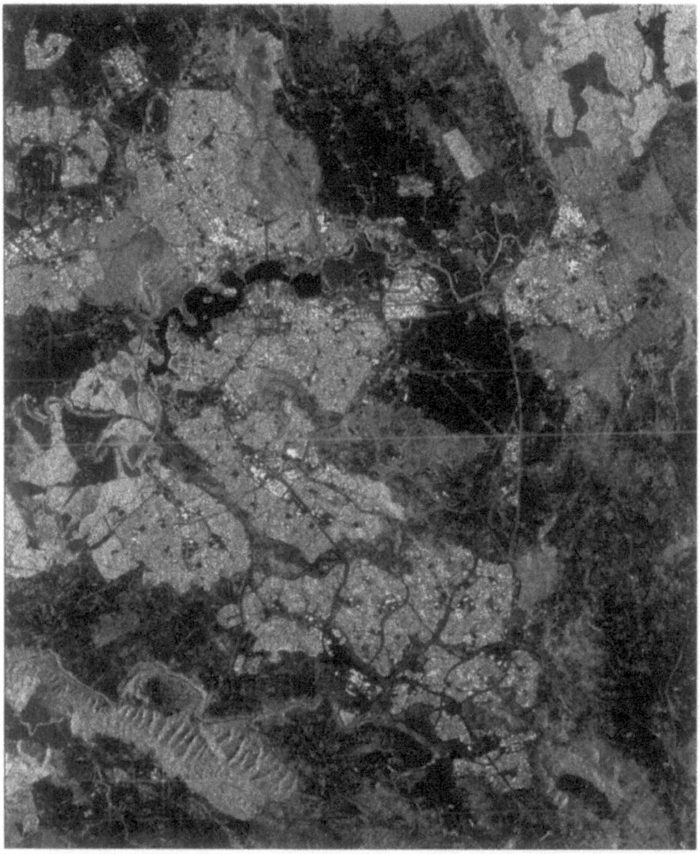

Figure 2-3. Canberra in the Australian Capital Territory region, photographed in April 1994 by the US space shuttle Endeavour. The area shown is 28kms high and 25 kms wide, approximately 30% of the ACT's total area. The Australian Parliament is visible on the shore of Lake Burley-Griffen in the upper right. Photo courtesy of NASA Visible Earth project.

4 POINTS TO REMEMBER

- The near shore, fine sediment environment of the North American Great Lakes, tributaries of the Fraser River in British Columbia, Canada, and upper Murrumbidgee River in Australian Capital Territory represent a variety of spatial extents, ecosystems and climates and are used as illustrative case studies.
- The case studies shared common elements in that each used collaborative setting of objectives, site selection criteria and standardized sampling and analysis.

Chapter 3

DEFINING THE REFERENCE CONDITION

Study Design and Data Collection

The Reference Condition Approach (RCA) to environmental assessment depends on the identification and sampling of reference sites. The result is a database of biological and environmental attributes that represent the Reference Condition. The database is used to describe variation in the Reference Condition and build predictive models that explain as much as possible of the variation in biological characteristics among reference sites by variation in environmental variables among the sites. The description of variation in Reference Condition, the development of predictive models and the ultimate assessment of test sites are discussed in Chapters 4 and 5. In this chapter, we describe and discuss the steps involved in designing a study that will define the Reference Condition.

Step 1: define the objectives of the study.
Step 2: determine the spatial and temporal extent and grain of the study.
Step 3: determine the criteria for acceptable reference sites.
Step 4: determine the appropriate location and number of reference sites.
Step 5: choose descriptors of the benthic invertebrate community and its environment.
Step 6: determine a data storage and management system.
Step 7: institute quality assurance protocols.

1. DEFINE THE OBJECTIVES OF THE STUDY

Establishing the purpose of an assessment or monitoring study is a critical first step in the process of defining the Reference Condition. Most such studies involve, at various stages, a number of groups of individuals including resource regulators, biologists and managers. Without explicit definition and agreement about why a study is being conducted, it is almost inevitable that one of the groups will be dissatisfied with the outcome.

Barbour et al. (1999) identified the following study objectives for biological assessments:

- characterizing the existence and severity of impairment to the water resource,
- helping to identify sources and causes of impairment,
- evaluating the effectiveness of control actions and restoration activities,
- supporting use attainability studies and cumulative impact assessments,
- characterizing regional biotic attributes of Reference Conditions,
- regional assessment of land uses.

To these we add:

- large-scale political decisions or policy or resource allocation.

Bioassessment studies are generally of two types: 1) observational studies, where space and time are the main experimental variables or treatments, and 2) manipulative studies, usually involving some intervention by the experimenter to control one or more external factors relative to the experimental units (Hurlbert 1984). The first are most common for environmental assessments such as the RCA, although experimental studies may be needed to verify the response of various indicators to environmental change.

Studies may be geopolitical and regionally based for the general purpose of biological assessment. In such cases, the study design would be as broad as possible in its coverage, resulting in "state of the region" reports. Examples are the Fraser River Action Plan reports (Gray and Tuominen 1998), Australia's Assessment of River Condition (Norris et al. 2001), or the UK five-year river surveys (Wright 1995). Alternatively, studies may focus on a single management issue, such as sediment contamination in lentic systems, resulting in a narrower focus of sampling and a much more specific study output (such as the near shore Great Lakes sediment study, Bailey et al. 1995, Reynoldson et al. 1995). Some monitoring applications address the day-to-day nature of industrial or residential effluents, thereby requiring a sensitive early warning component, while others may be designed for ongoing regulation of a particular industrial sector.

As we discuss in subsequent sections of this chapter, the reasons for conducting a study and using its results will influence how, where and what data are collected. The key to a successful and useful outcome is recognition of how the specific goal(s) of the study dictate both the general features and the details of the design.

2. DETERMINE THE SPATIAL EXTENT, TIME-SCALE AND GRAIN-SIZE OF THE STUDY

The second step in the study design will be influenced by geographic, physical or institutional factors, or some combination of these. There are several issues that must be addressed, including the geographic boundary of the study (the spatial extent), the type of habitats or water bodies in which the benthic community is sampled (the "ecospatial" extent), and the size of an individual site (the spatial grain-size). Similar decisions must be made regarding the temporal grain of the study. Both the temporal extent (total time for the study), and the "ecotemporal" extent (the particular period in which sampling will take place; e.g., early autumn) of the study must be considered.

2.1 Geographic boundary — spatial extent of the study

Definition of the geographic boundary of the study area is usually a straightforward decision determined by the study objectives. The limits may be defined by drainage basin or sub-basin, or by political or environmental boundaries.

For most studies, whether at large scale (state, provincial), local scale (regional, municipal) or point-source scale, the watershed (catchment boundary) of the drainage basin that encompasses the area of interest is a logical study boundary. There is at least a potential for ecological connection among sites within the one drainage basin (Figure 3-1). However, sites on either side of a watershed, or in similar situations in different catchments, may have more in common than sites within the same basin (see Newall and Wells 2000). The final boundaries can be set after considering later sections of this chapter. If a drainage basin extends beyond a politically defined study boundary, whether national, state, provincial or municipal, effort should be made to extend the boundaries of the study into the adjacent political jurisdictions so that as much as possible of the drainage basin is included.

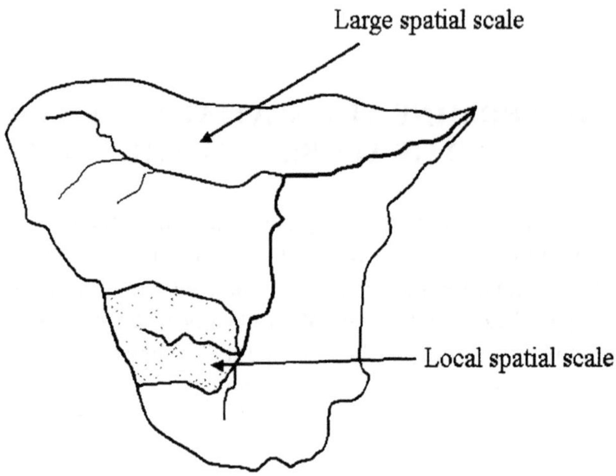

Figure 3-1. Defining the spatial extent of a study by the watershed

2.2 Habitat types considered — ecospatial extent of the study

Both the objectives of the study and the physical geography within the study area will affect the decision of the types of habitat to be sampled. Here we are largely concerned with the habitat or habitats in which the benthic community is sampled, since the environment of the invertebrate community will usually include descriptors well beyond the boundaries of the habitat itself within a site (catchment area and land use proportions, for example).

At the most basic level, the nature of the study usually dictates that a particular type of aquatic ecosystem (e.g., stream, lake, wetland, estuary) will be targeted, although some national programs, such as the Environmental Effects Monitoring (EEM) program in Canada, have a mandate to assess aquatic systems wherever effluent of a given type (e.g., from pulp and paper mills) is released. In the EEM program, near shore marine, lentic, and relatively large stream systems are all part of the mix of aquatic habitats that get sampled. If the study is dedicated to a particular type of aquatic habitat, such as streams (as in the Fraser River Basin or Australian Capital Territory studies) or lakes (as in our North American Great Lakes case study), as described in Chapter 2, there is still a need to define the types of habitats that will be sampled within these systems.

Whatever the choices at either the large or the small scale for sampling the biota, a description of the environment at a site will often include descriptions of other ecosystems that affect the local habitat. For example, a study of the littoral community in temperate lakes might include information about adjacent wetlands and inflow streams as part of a complete description of the environment of one lake's littoral zone.

Generally, one of two approaches is taken when sampling to deal with intra-site habitat variability (Figure 3-2; see also Parsons and Norris 1996). One approach is to sample all the major habitat types, allocating effort in proportion to the occurrence of each (about 10% riffle, 10% overhanging vegetation, and 80% run in Figure 3-2). The samples can be either composited or treated separately. Alternatively, a single, common habitat type (say, riffle in streams, or fine-grained sediment in near shore areas of lakes) can be sampled at all of the sites. A difficulty with this second strategy is that there may be substantial variability among the sites in the amount of this "common" habitat type that is available for sampling. Then, variation among the biota present at the sites could be at least partially a function of variation in the amount of habitat present.

Regardless of the approach used, we continually bear in mind when designing the study that the site is the unit of observation. Any sub-sampling done at the site is purely for the purposes of getting a more comprehensive sample or for characterizing spatial variability within the site as part of its description. We must leave behind the notion that points or areas sampled within a site are in any sense "replicates", whether used as such or for consideration of possible data transformations (Norris et al. 1996).

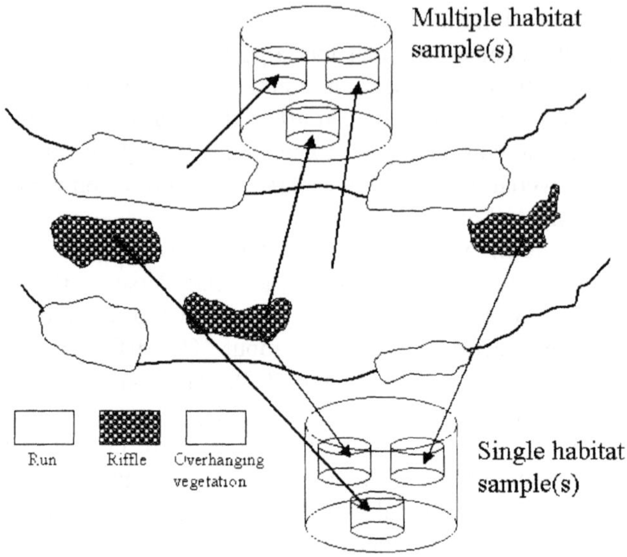

Figure 3-2. Two strategies of dealing with intra-site variability in the habitats sampled

How sampling is performed and whether particular habitats are sampled will depend on the distribution of habitat types among test sites and the objectives of the study. For example, an initial objective of the RIVPACS program in the UK was identification of taxonomically rich sites potentially worthy of protection (Wright et al. 1984). Thus, the aim of sampling in the RIVPACS program was to maximize the taxonomic list for a site, and so all habitats were sampled in proportion to their occurrence at a site within an overall maximum time. In the National River Health Program in Australia, on the other hand, the aims were (and are) to minimize variability to facilitate comparisons (Parsons and Norris 1996).

Two main factors dictated the choice of habitat to be sampled in Australia: first, the likely availability of the habitat throughout the study region; and second, the habitats most likely to respond to possible stressors. In all Australian states and territories, two habitats are sampled to ensure test sites will be covered, thus providing more confidence in the assessments (Simpson and Norris 2000: see multiple assessments in Chapter 6). Different habitats are sampled in different Australian states and territories: for example, riffles and edges in the Australian Capital Territory, main channel and macrophyte beds in South Australia, and edges and sand bars in the

Northern Territory (crocodiles preclude personnel entering the channel). The habitats sampled and the methods are consistent within a state or territory and for the model used to analyze the data. Therefore, the outputs and assessments are equivalent for the whole country.

2.3 What is a site? The spatial grain-size of the study

When using the RCA for biological monitoring and assessment, we take the site as the unit of sampling — a fundamental shift from traditional sampling practice. If we wish to characterize the variability *among* reference sites to assess whether or not a particular test site is in Reference Condition, it makes no sense to compare the mean of several observations in one reference site to the mean of observations from one test site. Given the premise of the site as the unit of sampling, we must, as in the case of the study area as a whole, carefully define the boundary of an individual site. The most critical issue at this end of the spatial scale is the desirability of collecting data from sites that are far enough apart to be ecologically independent, or at most minimally correlated with one another.

In stream studies, site length is usually 5–10 times the average bank-full width of the stream (Newbury 1984). Rabeni et al. (1999) demonstrated that a single stream reach (~20 x mean bank width) is adequate for characterizing reference streams, and showed that sampling additional reaches did not substantially improve the estimation of community descriptors. If there are sufficient streams within the sampling area, then only one site need be established per stream, although this becomes progressively more difficult with higher order streams. Sites should be located away from the influence of tributary streams or other water bodies. If a single stream is sampled at more than one site, the sites should be separated longitudinally by a minimum of 100 stream widths or about 10 pool–riffle sequences. So in a stream with an average width of 10 m along its length, sites will be 50–100 m long, and there will be a minimum of 0.5–1 km distance between adjacent sites on the same stream. However, these are just "rules of thumb". Although the lack of independence of sites along the length of a stream has long been recognized (e.g., Eberhardt 1976, 1978, Eberhardt and Thomas 1991; Hurlbert 1984, Norris and Georges. 1993), there has been remarkably little work done to quantify the degree of longitudinal, spatial correlation in streams.

Sites within lakes and ponds, or in lakes that are along a stream network, are less obviously linked than sites along a stream. However, these systems also have at least the potential for spatial correlation, since they reside in the same catchment area. For lake studies (see EPA 1998), we suggest that a minimum site area is 1 ha (100 m x 100 m). If the lake consists of a single basin of less than 10 ha surface area (Figure 3-3a), then the entire lake can be considered the site. If the lake has more than one major basin, each less than 10 ha, then a separate site can be established in each basin (Figure 3-3b). If the lake or its individual basins are larger than 10 ha, then there more than one 10 ha site could be established in each major basin, as long as they are separated by more than 1 km (Figure 3-3c).

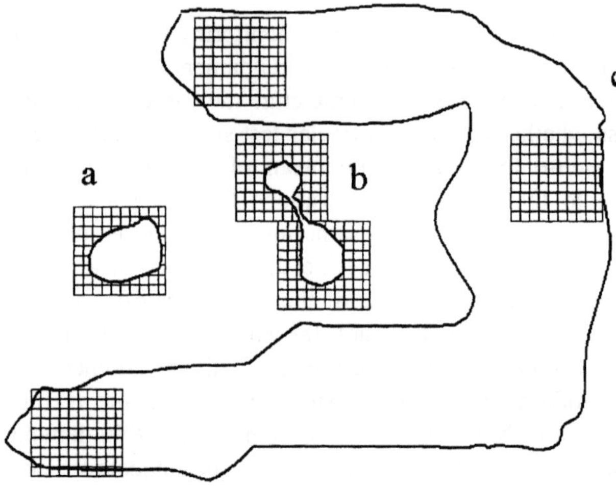

Figure 3-3. What is a site? The lentic environment. (a) Single basin, small (<10 ha) lake. (b) Two, small (<10 ha) basins in a lake. (c) Three sites in a large (>10 ha) lake.

2.4 The temporal scale of the study

There are two issues related to choosing the temporal scale of the study. The first relates to seasonal differences in the environment, which are known to affect aquatic organisms (Lemly 1997), while the second relates to longer, year-to-year changes.

There are various opinions on the best time of year to collect information about benthic invertebrate communities. In temperate climates, the autumn may be best in both lotic and lentic systems, because most of the invertebrates that spend part of their life cycles out of the aquatic environment (such as Insecta) are aquatic immatures during the autumn (Merritt and Cummins 1984). Additionally, environmental conditions and weather are often more predictable in the autumn, thus making the logistics of sampling simpler, and reducing the variability associated with the description of the site. However, the invertebrates are at a younger stage of development during this season, making identification more difficult. A further, more practical, difficulty could be a lack of students and volunteers, who may only be available during summer periods. Ideally, multi-season sampling of the same site would be conducted, since bioassessments have been shown to vary significantly among seasons (see Reese et al. 2001).

When data are collected in multiple seasons, either separate seasonal or combined multi-season models can be constructed. In the first UK models, Furse et al. (1984) showed that combined models were more accurate for site prediction than separate models. However, more recently, Linke et al. (1999), who included season in their sampling, were able to increase the predictive power for detecting degradation and found that separate predictive models for each sampling season gave the best results. Both the British RIVPACS and Australian AUSRIVAS employ discrete season sampling, and predictive models are constructed to accommodate the seasons. RIVPACS has models for each of three seasons, each combination of two seasons and all seasons together (Wright 1995). AUSRIVAS has models for each of two seasons and a combined season model. The "temporal grain" of sampling could be within one season, or across several seasons. The time-scale chosen will have implications for study design, and the potential use of the models must be considered in the study design phase.

Annual changes in Reference Condition are less directional and predictable than seasonal variability and are influenced by large-scale phenomena such as floods and drought. To date, there have been few aquatic studies that offer a long time-series of observations related to environmental conditions. The longest period over which reference site data have been collected and used in predictive models occurs with RIVPACS, where progressively larger datasets have been assembled and analyzed over 15 years. The strong relationship found there between the macroinvertebrate fauna at reference sites and site environmental attributes implies that assemblages are reasonably stable over time. However, it would be unwise to be complacent about variation in community structure over time, particularly in geographical areas subject to climatic extremes and in small streams that are subject to chance events that have long-term effects on their resident communities (Wright 1995). At a recent workshop on predictive modelling (Wright et al. 2000), no conclusion was drawn about the validity of models constructed with data collected in a narrow temporal window. Repeated sampling of a sub-set of reference sites should be incorporated into all programs, to examine change through time (see example, Chapter 6). The timeframe of a study, "how long it lasts", is primarily dependent on its funding and mandate, although natural variability should be considered within this context (drought/flood cycle, freeze/thaw cycles). Most bioassessment or biomonitoring studies, even if at a local scale, are conducted over several years.

Finally, we again emphasize that all these suggestions for the spatial and temporal grain and extent of the study address are for use when the benthic community are the biota of interest. They would be different for other communities. The environmental predictors of the community structure at any spatial scale may need to include information at smaller scale (such as the vertical and horizontal spatial variability of substrate size within the site) and larger scale (say, annual mean precipitation). In fact for several variables (such as temperature or dissolved oxygen), instantaneous measures are of little value and some sort of picture over a longer time is required (say, monthly or annual degree days, see Marchant et al. 1997).

3. DETERMINE THE CRITERIA FOR ACCEPTABLE REFERENCE SITES

Once the objectives scale of the study have been clearly defined, it is necessary to establish criteria for selecting reference sites. This important part of the study will define what is "good" — a definable goal for ecosystems that perhaps have been degraded. Although some "ground truthing" of candidate reference sites is carried out by visiting the sites and collecting data from them, it is important that criteria defining potential reference sites are established in advance of the actual sampling, and without regard to the invertebrate community found at a site. Put simply, the structure of the community itself is not used to identify the site as reference or otherwise. It is the variation among communities in Reference Condition that we are hoping to capture in our sampling of both communities and their environments, and that will describe the characteristics of reference sites.

It is not always possible to find pristine reference sites; therefore criteria such as "never exposed to human disturbance" are unproductively constraining. A degree of pragmatism needs to be included in any definition of the Reference Condition. Many areas with a long history of human modification can have suitable reference sites, in spite of, say, the effects of grazing activity in the highlands of Scotland, or the use of fire by indigenous peoples in North America and northern Australia. The goal when establishing criteria is really to define explicitly the reference or acceptable ''healthy'' ecosystem for the context, so that there can be common decision rules emphasizing expectations of *minimal exposure to the stressors of concern*. The notion often applied to Reference Condition is that of "best condition that could be expected" (Reynoldson et al. 1997). This notion of Reference

Condition is central to currently accepted ideas of "biocriteria" being developed by the USEPA (Davis and Simon 1995).

There will inevitably be situations where there are limitations to the application of the RCA, and these are likely to be most often seen in lowland rivers and large lakes that are few in number and unlikely to have undeveloped equivalents for comparison. (See Marchant et al. 1995, Reynoldson et al. 1995, Thoms et al. 1999, Norris and Thoms 1999; and see Chapter 6 for alternative assessments.) Lowland rivers and large lakes usually have a long history of intense flow regulation, agricultural development and pollution (Petts 1989), and hence Reference Condition may be hard to obtain from a study of the present-day ecosystems. Other methods such as paleo-ecology may be appropriate for establishing Reference Condition there (Thoms et al. 1999).

We base our definition of acceptable Reference Conditions on ecological understanding of a system, and, as already stated, criteria establishing acceptable reference sites should be set *a priori*. As indicated in Chapter 4, analysis of community and environmental data from sites that satisfy the criteria for Reference Condition may reveal some unexpected conditions at specific sites, causing them to be deleted from the reference set. It may be only at the stage of field sampling that obvious effects of human disturbance on the site (say, sedimentation of fine material from urban runoff) are revealed. Sites that were thought to be reference sites may need to be re-categorized as test sites at this stage.

The selection process adopted for reference sites (in rivers) in Australia's national program is described by Davies (1994), but can be generalized for any kind of water body (Figure 3-4, Table 3-1). The procedure specifies that reference sites should cover a wide range of physical and chemical conditions encompassing all types of river discharge, size and altitude within the study area. Reference sites are chosen on the basis that they are minimally disturbed, acknowledging that lowland rivers are expected to be affected by a range of land use practices. Reference sites are required to be representative of the reach or location and not subject to significant flow regulation, and they should be accessible and safe during sampling. Selection of reference locations, and sites within them, is an iterative procedure, although criteria should be stated up-front, *a priori*. The flow chart for a generalized procedure (Figure 3-4), which has been adapted from Australia's National River Health Program (Davies 1994), identifies the characteristics that can contribute to site selection (Table 3-1).

In situations where enough near-pristine locations are available to generate a relevant and sufficiently large database, these locations should be used preferentially. If near-pristine locations are not available to generate a database of reference sites that encompasses the environmental variability of the region, then a database that includes sites that have received less intense impacts might justifiably be used. For example, sites affected by land clearing, moderate flow regulation, or non-intensive agriculture or forestry might be judged to be relevant "least disturbed" reference sites, or long-term mitigation 'goals', against which to assess mitigation of gross impacts (Davies 1994).

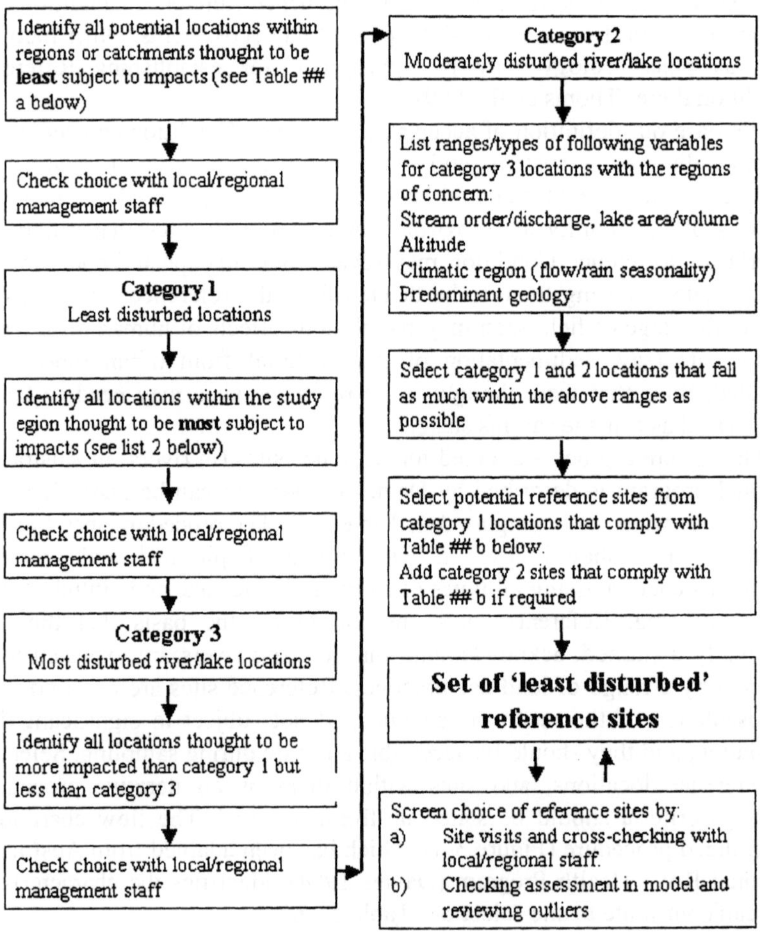

Figure 3-4. Generalized flow chart for selecting reference sites. Adapted from Davies (1994).

Table 3-1. Criteria for reference location and site selection (from Davies 1994)

Characteristic	Description
(a) *Types and characteristics of impacts used in identifying "least disturbed" locations and sites*	
Major water impoundments, extractions or diversions	Presence of changes to flow regime or water balance and/or water quality such as large, repeated short-term fluctuations in discharge; suppression of medium to large-scale natural discharge fluctuations; significant flow reduction or dewatering; significant changes in temperature, dissolved oxygen or other water quality variables; sediment load from construction or increased bed and bank erosion; release of toxic algae or chemicals used in water treatment. **Impacts** on stream hydrology/lake water balance, sediment load, temperature, water chemistry
Vegetation clearance	A percentage of land cleared, time since majority of clearing activity, presence of active clearing **Impacts** on stream hydrology/lake water balance, sediment load quality, temperature, water chemistry (including nutrients)
Catchment urbanization	A percentage of catchment urbanized, and potential source of storm water **Impacts** on stream hydrology/lake water balance, sediment load and quality, water chemistry, contamination
Presence of roads, tracks or service corridors	Number and type of stream crossings, lake access points, proximity, management of road/access points crossing drainage, recreational activities **Impacts** on stream hydrology/lake water balance, sediment load and quality, impacts of recreational activities
Channel and bottom modification	Degree of channelization, channel reforming, dredging, infilling, time since last major works performed **Impacts** on stream hydrology/lake water balance, sediment load, substrate composition, bed stability
Bank and shoreline degradation	Lack of bank and shoreline vegetation, extent and activity of eroding bank or shoreline. **Impacts** on bank/shoreline and bed stability, sediment load, substrate composition, water temperature, invertebrate food resources
Floodplain and wetland drainage	Presence of intensive drainage of floodplains and catchment, and associated wetlands, proximity to site **Impacts** on stream hydrology/lake water balance, sediment load, water chemistry
Forestry	Percent of catchment logged, predominant width of buffer strips, if used extensively (>30 m desirable to reduce impacts), presence of active logging (in the last ≤ 5 years or less) logging **Impacts** on stream hydrology/lake water balance, sediment load and quality, temperature, water chemistry, habitat structure (wood debris)
Mining or extractive industry	Presence of mine or related industry in the site's catchment, its potential as a source of pollutants (release points of waste water and its treatment); proximity to site, percent area of catchment disturbed **Impacts** on stream hydrology/lake water balance, sediment load and quality, chemical contamination

(continued next page)

Pollution sources	Point-source discharges such as sewage treatment plants, light and heavy industry (e.g. e.g., chemical plants, tanneries), agro-industry (e.g. e.g., piggeries, dairies, food processing works), mines and mine-processing works, storm water. Diffuse -source releases such as leachates from tips and chemical storages leachates, contaminated groundwater, saline surface and ground waters, pesticide runoff. **Impacts** on water chemistry, sediment quality, organic load and chemical contamination
Intensive agriculture	Percent of catchment developed for intensive agriculture, types of crops grown, sources and releases of waters used for irrigation, proximity of cropped areas to streams; routing of flood waters through cropped areas, use of pesticides and herbicides, use of fertilizers **Impacts** on stream hydrology/lake water balance, sediment load and quality, temperature, water chemistry, contamination
Grazing	Percent of catchment cleared for grazing, number and control of stock water access points and stock routes, use of fertilizers, water abstraction for stock watering **Impacts** on stream hydrology/lake water balance, sediment load and quality, temperature, water chemistry

(b) Discrimination for selecting the positions of reference sites within a location

Site features		Requirements
The site should	a.	be deemed representative of the major characteristics of the overall location
	b.	be subject to no, or as little as possible, human-related disturbance
	c.	be relatively accessible and safe during sampling operations
	d.	be sufficiently distant from any pollution source for recovery to take place in the biological community composition, e.g., ≥ 10 km for small streams, separate lakes if area ≤ 5 km^2, ≥ 20 km for large rivers, ≥ 5 km in large lakes dependent on circulation
The site should not	a.	lie within 50 m upstream, or 500 m downstream, of a traverse or ford; dam, weir or waterfall >5 m; livestock watering area; significant confluence or discharge or lake inflow; significant water diversion/abstraction; areas subject to channelization, dredging or weed removal
	b.	be subject to significant adverse water regulation (large abstractions or releases that severely modify water quality, temperature or discharge)
	c.	be near source(s) of identifiable pollution (see above)

4. DETERMINE THE APPROPRIATE NUMBER AND LOCATIONS OF REFERENCE SITES

One of the most difficult stages of designing an RCA study is deciding how many reference sites to sample and where they should be. Although from the point of view of sampling theory (e.g., Cochran 1977), the task of "putting the points on a map" is relatively simple, the notion that the natural variability in a large geographic area can be adequately characterized with relatively few reference sites (e.g., 50–500) is often difficult to convey to those most familiar with the area and its heterogeneity.

The required number of reference sites is dependent on many factors. For example, geologically complex areas are likely to need more reference sites on their lakes, streams and rivers than a geologically uniform land mass of similar area. Also, many people have strongly entrenched notions about how to stratify sampling effort within the study area. For example, ecoregions (Omernik 1987), or some variation on ecoregions, are often considered a necessary geographic basis for stratification of sampling effort. It is important to remember, though, that the task of the RCA is to characterize variation among the biota in reference sites across the entire area of the study, and later to use variables that describe the environment of each site (including, possibly, its ecoregion) to explain this variation in a predictive model. The important focus at the design stage is to adequately encompass the variation among biota within the geographic area. The particular stratification strategy used should be in aid of this "complete coverage" goal (see Hawkins et al. (2000) for a discussion).

Although the number of sites to be distributed among the strata will be influenced by the budget for the project, the desired level of precision in the predictive model must remain an important consideration. The success of the prediction system is also heavily dependent on the strength of the relationship between the biological and environmental attributes of the reference sites, and on identification of a relatively small set of environmental attributes with high predictive capability.

As with any sampling procedure, the more observations (sites) collected, the more precise will be our characterization of variation among the communities and the relationship between the community and its environment. There is, as always in sampling a population, a law of diminishing returns — the effort and study resources for thousands of sites are rarely immediately attainable. However, large databases are becoming more common. For example, Wright (1995) reports almost 30,000 sampling occasions for 9000 sites for a five-yearly assessment of rivers in Britain. Australia's National River Health Program and First National Assessment of River Health sampled 6000 sites [across the whole of Australia] on 2–4 occasions. Several American studies (such as Klemm et al. 2002, Hawkins et al. 2000) also report on programs where hundreds of reference sites have been sampled.

It is perhaps more relevant to consider the *minimum* number of reference sites necessary. We consider 25 to be the minimum necessary for even a pilot project. With 25 sites, one can make at least a crude estimate of variability among the reference communities in structure and composition, and derive an indication of the important environmental variables that explain this community variation, via statistical model building (as described in Chapter 4). If there is relatively little variation among the reference communities, or if most of this variation is explained by the predictive model, then it may be possible to use even this small a dataset for the actual monitoring and assessment of test sites (see Chapter 5). An over-riding note here is that there is little sense in including reference sites for habitats that are not found at the test sites. For example, reference sites that adequately describe the variation in upland streams, small lakes or ponds will be of little value if all the test sites are in lowland sections or large lakes — a problem commonly encountered because much human activity occurs in relatively flat lowland and coastal areas. In the end it is very difficult to specify exact numbers of sites: the numbers are primarily determined by the spatial complexity of the study area, the resources available and the desired certainty of the assessment.

As mentioned, allocation of the reference sites to specific points on the map usually includes some sort of stratification so as to better sample and characterize the total variation among reference communities within the study area. If sites are located by simple random sampling of the habitat within the geographic area, there will be a bias towards the more geographically common areas on the map. This is not necessarily desirable. If we want to characterize variation among lentic sites, should our selection of sites be biased towards the larger lakes in the study area? Similarly with stream studies, if reference sites are chosen at random along the total length of streams in the area, should there be more sites along the dominant stream order. Stratification by stream order, and equal sampling effort in each order will ensure better characterization of variation among reference sites in the stream system (Reynoldson et al. 1997).

Whichever scheme is selected for assigning reference site locations, it should be directed towards covering the range of natural conditions within the study area. The future use of the reference database is not known and one of the advantages of the RCA is that the reference site data can be re-used. Therefore, the database and predictive models need to be developed to ensure that they have the maximum potential applicability in the future.

5. CHOOSE DESCRIPTORS OF THE BENTHIC INVERTEBRATE COMMUNITY AND ITS ENVIRONMENT

The criteria for reference sites have been established, the spatial and temporal extent and grain of the study have been defined, and sites have been located on a map. Now, both the biological community and its environment at each site need to be described so that a predictive model for the reference community can be constructed. The model will enable us to determine the reference community to be expected at a test site, given the environmental conditions at the site. For the benthic macroinvertebrate community, the data obtained will be the number of individuals (or biomass) of each taxon in the community (see 3.5.1). We discuss further treatment of this "fundamental dataset" in the next section. It is less clear what should, or needs to be measured to adequately describe the environment.

We have not attempted to describe the various methods used to collect data because there are many volumes dealing with this aspect of bioassessment programs (e.g., Hynes 1970, Hellawell 1986, Barbour et al. 1999). Appropriate and consistent methods should be chosen. Regardless of the decisions concerning sampling methods and habitats, it is critical that the *sampling effort is equivalent at all sites*. Many of the biological features of a site will be related to sampling effort, such as number of taxa, rare taxa, and abundance estimates. There has been considerable debate in the scientific literature about areas of collection, taxonomic resolution (Cao et al. 1998, 2001 Karr and Chu 1999, Marchant 1999, Cao 1999), sub-sample size or whole sample sorting and identification (Barbour and Gerritsen 1996, Courtemanch 1996, Vinson and Hawkins 1996, Sovell and Vondracek 1999, Carter and Resh 2001), and taxonomic groups used (Hawkins and Norris 2000, Rabeni and Wang 2001).

Whatever collections are made, they will only ever be a *sample* that, it is hoped, represents the site. The purpose of using the RCA, and indeed most environmental assessment, is to determine the status of test sites, through comparison with reference sites. Therefore, it is important that sampling effort is equivalent in reference and test sites so that comparisons between them are valid.

The result of an appropriately designed sampling program will be the acquisition of what we call the "fundamental dataset" of environmental assessment.

5.1 The fundamental dataset

We refer to the set of descriptors (p taxa and q environmental descriptors) actually measured and recorded for each of n reference sites (Figure 3-6) as the "fundamental dataset" of the assessment. Some of these descriptors may be descriptive statistics from data aggregated across a spatial or temporal scale (e.g., mean Taxon Richness per point observation at a site, standard deviation in annual mean water temperature among years). The benthic community portion of the fundamental dataset for a given site is represented by the abundance (or biomass) density (x_i) of each of the p taxa observed at any of the n sites, their proportion of the total abundance or biomass (p_i), or just an indication of their presence or absence (b_i).

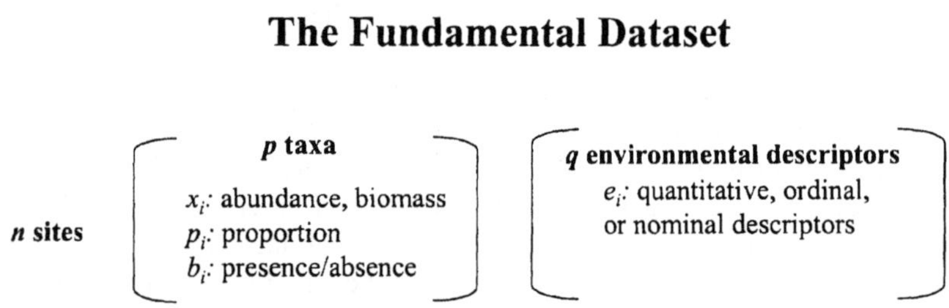

Figure 3-5. The structure of the fundamental dataset in RCA assessment studies.

The environmental descriptors component of the fundamental dataset for a given site is the value for each of the q environmental descriptors (e_i), which may be continuous (e.g., water depth), proportional (e.g., % sand substrata), ordinal (e.g., habitat assessment score), or nominal (e.g., primary land use in catchment). It is important to consider environmental descriptors at a variety of spatial and temporal scales to adequately characterize the site.

Benthologists have not traditionally made much use of large-scale descriptors of the catchment area (cf. Hunsaker and Levine 1995). Such variables can be useful for both general description of the study area and predictive modeling. In particular, many potentially valuable predictors of the chosen biological descriptors (e.g., catchment area upstream of a stream site or catchment geology) may be measured "from a map" and often have the strongest relationships with biological descriptors at reference sites (Wright 1995, Norris and Hawkins 2000). These large-scale variables may therefore be very useful in building predictive models of variation in the biota. It is important that we extend our thinking beyond only descriptors that can be observed and measured in the field.

Table 3-2. Examples of environmental descriptors in the fundamental dataset

Scale	Descriptor
Study area	Site catchment as a proportion of entire study area
	Proximity of site catchment to other site catchments, shared edges
Site catchment area	Morphology (area, perimeter, depth, volume)
	Drainage density
	Land cover (including vegetation)
	Surficial and bedrock geology
	Ecoregion
	Climate: annual mean, intra-, inter-annual variation
Site	Location (latitude, longitude, altitude, stream order)
	Riparian (bank) vegetation
	Aquatic primary producers
	Channel and shoreline structure
	Water turnover (discharge in stream, turnover time in lake
	Habitats (e.g., riffle, macrophyte bed)
	– total, relative area
	– habitat patches
	– number
	– mean area
	– mean perimeter
	– isolation
	Water quality: alkalinity, conductivity
	Physical habitat
	Catchment area upstream from the site
	Distance from the source
	Stream order
	Slope, aspect

5.2 Descriptors derived from the Fundamental Dataset

There are two types of descriptors that can be derived from the fundamental dataset: (i) internal descriptors that are derived directly from the fundamental dataset, and (ii) external descriptors that use independently collected information (e.g., pollution sensitivity of invertebrate taxa) along with the fundamental dataset.

5.2.1 Internal descriptors

In many cases, either the individual internal descriptors themselves or functions of a group of descriptors can adequately describe the variability in the community or its environment among reference sites. It is quite common to describe biotic variation using multivariate techniques applied to x_i, p_i, or b_i values from the fundamental dataset (see Chapter 4). Additionally, summary descriptors of the community, such as richness or diversity, are easy to calculate and may clearly describe major trends of variation among reference communities. Simple structural descriptors of a community such as richness, Simpson's diversity (Simpson 1949) or Shannon's diversity (Shannon 1948) — essentially the inverse of the variance of taxon proportions in the community — or equitability or similarity measures, and other such indices (see Washington 1984 and Barbour et al. 1999) are useful as synthesis descriptors of communities. However, there has been consistent criticism highlighting the limitations of some such summary descriptors, particularly diversity indices (e.g., Hurlbert 1971, Goodman 1975; Washington 1984; see Chapter 1). In analyzing our case studies (Chapters 4 and 5), we use Taxon Richness (number of taxa) and MCDist to illustrate single descriptors of the biota. Taxon Richness is probably the most dependably estimated measure of diversity of a community (Green 1979). The less familiar MCDist is derived from the Percent Model Affinity measure of Novak and Bode (1992). It assesses, taxon by taxon, the biota at a particular site in comparison to the typical or median community, determined by calculating the median relative abundance of each taxon across all the reference sites. The differences between the relative abundance of each taxon and its respective median are aggregated in the Bray–Curtis distance measure, which we call MCDist. If the biota at a site is exactly the same as the median community, MCDist is equal to zero. The more distinct a site's biota is from the median community, the closer the MCDist will be to its maximum value of one.

Environmental descriptors (e_i), whether quantitative or categorical, are most often analyzed directly in the context of predictive models or other statistical techniques. However, there may be value in summarizing such information using descriptive tools analogous to those used with community data. Thus, the "richness" of the substrate habitat as indicated by the number of substrate particle size categories present at a site, or the variety of land-use patches in the drainage basin, may indicate important effects on the biota found at a given reference site.

5.2.2 External descriptors

With both community and environment data, it is often useful to combine additional, independently collected information with that collected at the sites themselves to more completely describe variation among reference sites. For example, rather than classifying the members of the benthic macroinvertebrate community taxonomically (e.g., at family or genus level), they might be classified on the basis of their sensitivity to organic pollution (e.g., Hilsenhoff 1988) or their functional feeding group (Cummins and Klug

1979). The numbers of individuals in each of these groups might then be subjected to either multivariate analysis or calculation of some summary index.

The environment of the site can be subjected to a classification system that combines conditions of the site's environment with some independently derived scale of habitat quality. The USEPA Habitat Assessment (Barbour et al. 1999) is an example of a description of the environment that combines basic environmental descriptors (e.g., substrata, flow rate in streams) with independent information about the "value" of different habitat conditions for the benthic macroinvertebrate community. In addition, external information about presumed safe levels of various elements (e.g., Cd concentration) can be combined with measured values of elements at the reference sites to calculate external descriptors of a site.

6. DESIGN A DATA STORAGE AND MANAGEMENT SYSTEM

A significant problem related to research involving the collection and analysis of ecological assessment data is the maintenance of information. Every research study involving large amounts of data (best represented as $n \times p$) requires a carefully planned and managed data storage system. This is certainly true of RCA studies, where often hundreds of benthic macroinvertebrate taxa and tens of environmental descriptors are recorded for each reference and test site. Because of the nature of the data, most studies presently use some form of relational database. This is much more efficient, in both storage and data entry, than common alternatives such as spreadsheets. Such a method of information storage and retrieval is vital for handling large, complex datasets, while maintaining efficiency and accuracy for data entry and retrieval.

With the introduction of personal computers and automated data collection techniques to scientific study, the quantity and complexity of the data gathered has grown exponentially in recent history. The volume of data that must be entered, stored, and managed in some form of database has led to several significant issues (Hernandez 1997, Silberschatz et al. 1997) that must be addressed by database managers if the information is to be of use:

Data redundancy — traditional row–column data storage formats can lead to significant redundancies. This can make data difficult to sort through, and cause significant storage problems.

Inconsistency — as datasets mature, and more people become involved in their management and use, inconsistencies can emerge because of a lack of standards.

Access difficulties — the database system must be able to provide easy access to the information stored within it, if it is to have any value.

Data isolation — to simplify storage, it is tempting to break data into smaller files, to reduce file size. Unfortunately, this can lead to files being stored in various file formats and types, which may or may not be compatible for future use.

Data integrity — information must conform to *consistency constraints*. However, as situations change, these constraints may need to be modified. Rigid systems can limit the ability of the user to make these changes, making the data less applicable.

7. ENSURE DATA QUALITY

Every aspect of an RCA bioassessment must be subjected to quality assurance and control. In any project, there are several easily identifiable stages where steps can be taken to avoid future data quality problems. These are in the study design, training, sample collection, sample processing and data entry.

Errors that occur in the first stages of sample collection, field measurement, and sample preservation and transport may be serious and are usually irreparable. Quality assurance and control for data should have procedures prescribed that, when adhered to, provide confidence in the quality of the data. These initial stages are largely outside the ambit of this book.

Just as important are data quality protocols that minimize errors during laboratory processing of samples, data recording, and data transcription and entry. These procedures are described below in the section on exploratory analysis.

In the design phase, projects can be assessed by independent consultants or scientists. Decisions about basic issues, such as stratification and allocation of sampling effort among habitats, can be offered for criticism and adjusted accordingly. Incorporation of some replication to assess measurement variability can provide an error-term associated with the estimation of each variable. One approach is to take replicated samples at a proportion of sites (e.g., 10–20%), thus enabling an estimate of measurement error at the site scale.

Ensuring that operators have appropriate and sufficient training before field sampling can greatly assist in reducing variability between operators. This can be as formal as a training and certification process, or as simple as a consensus field day spent at one or more sites typical of those to be sampled. It could include repeated sampling at a site by different sampling teams.

However, some pre-sampling program to ensure consistency is essential. It greatly increases the consistency of field data collected by different people, and addresses the inevitable vague or ambiguous aspects of the field protocol. Everything, from the point and method of collection of water samples to the best way to clean benthic macroinvertebrate samples in the field before preservation, should be discussed in this pre-sampling program.

Quality assurance and control (QA/QC) are important at several points in the sample processing of the benthic macroinvertebrate community. Most bioassessment studies use some version of sub-sampling, although it is still somewhat controversial. The most popular approach to sub-sampling involves randomly picking a given proportion of the unsorted sample (often a fixed area on the bottom of a pan), and sorting the invertebrates from this sub-sample. The process is repeated until a minimum sub-sample size (usually between 100 and 300 organisms) has been sorted out. Whenever a new sub-sampling procedure is adopted, a small study should be conducted on completely processed samples, to make sure that the new method is unbiased and so that the accuracy of the sub-sampling technique can be documented.

The benthic invertebrates in the sub-sample are usually identified to some pre-determined taxonomic group, such as the lowest possible taxonomic level. This will vary somewhat among and within taxa because of variability in the detail of currently used taxonomic keys and variability in the life stage and development of the benthic macroinvertebrates present in a particular sample. Often more than one person is involved in sample sorting, counting and identification. Quality assurance steps need to be developed for both sorting and identification. To estimate sorting efficiency, samples should be retained. A proportion of the saved samples should be rechecked by another person to assess the recovery rate: a target of 95% recovery is the objective of many laboratories. Performance charts should be maintained by each laboratory.

Two simple measures can be initiated to ensure taxonomic data quality. First, all laboratories should maintain a reference or voucher collection of all organisms identified. This will allow all identifications to be verified by an independent taxonomist. Second a proportion of all samples should be re-identified independently, on an ongoing basis, allowing the laboratory to report a taxonomic error rate.

The final potential source of error in the fundamental dataset is in data entry. Data from field observations (e.g., percent shading of stream), data collected from geographic information systems (e.g., land cover proportions and absolute areas) and maps, the invertebrate counts and any other analyses will be entered into the database. Databases can incorporate error checking in the data-entry form (e.g., rejects percent stream is shaded over 100% or expressed to two decimal places), which avoids major data entry errors. Transcription errors can also occur, and a system of random periodic checking against field sheets can establish whether or not it is a substantial problem. Finally, all bench and field sheets should be retained, to allow for recovery of or verification against the original data.

8. CASE STUDIES: DEFINING THE REFERENCE CONDITION

8.1 Near shore environment of North American Great Lakes

8.1.1 Objectives

Until the early 1990s, decisions about sediment contamination in the near shore environment of the Great Lakes were based almost exclusively on chemical concentrations in the waters. However, there was widespread dissatisfaction with this chemical-only approach. Not only did it frequently fail to achieve its objectives (Cairns and Van der Schalie 1980, Long and Chapman 1985, Chapman 1986, 1990), but also it was so conservative that it had limited value in assisting management decisions (Zarull and Reynoldson 1992). The very large areas of sediments designated as ''contaminated'' using such chemical criteria made many remediation methods impractical (Painter 1992). It became evident that biological information was needed in addition to data on chemical concentration.

The purpose of environmental assessment and management is ultimately to promote the maintenance of biological integrity. Therefore sediment quality objectives should include biological targets together with chemical surrogates. This approach was the basis of the sediment quality triad proposed by Long and Chapman (1985) and strongly endorsed in two International Joint Commission (IJC) reviews (International Joint Commission 1987, 1988) of the assessment of sediment problems in the most contaminated areas of the Great Lakes.

Both the sediment quality triad and the IJC promoted laboratory and field biological assessment in identifying contaminated sediment. In both cases the use of invertebrate assemblage structure was suggested as the appropriate field component and toxicity testing as the laboratory component. The objective of the study was therefore to develop a set of numeric biological targets for assessing sediment contamination, by:

- establishing an array of reference sites in areas of fine grained sediment in the near shore areas of the Great Lakes
- identifying benthic invertebrate community assemblages at those reference sites
- determining the normal response range in a suite of invertebrate test organisms in laboratory bioassays
- establishing the feasibility of using predictive modeling to set numeric targets for identifying effects related to sediment associated contaminants.

8.1.2 Spatial and Temporal Extent and Grain-size

Samples were taken over the 1991 1993 study period from the near shore zone of the Great Lakes using a 0.25 m^2 box corer. The minimum distance between sites was set at 100 m. Most sites were sampled in late summer or early fall over a three-year period. The sampling period ranged from mid-July to the end of October, the median sampling date was 12 September, and 80% of the sites were sampled in August and September. In addition, some sites were sampled in more than one of the field years and four sites were sampled monthly over two years. These data allowed the effects of both annual and seasonal variation to be determined.

Table 3-3 Study design stage decisions for the Great Lakes case study

Study Design Stage	Decision
Spatial extent	North American Great Lakes, near shore (up to 10 km) fine grained sediment
Spatial ggrain	Box core ($0.25m^2$) and environs ($\sim 100m^2$)
Temporal extent	3 years, summer–autumn, 5–10% of original sites repeated annually
Temporal grain	1 hour site visit
Criteria for Reference	More than 10 km from a point source of pollution No agricultural or urban shoreline development
Criteria for rejection as Reference	Complete absence of invertebrates More than 50% mortality in one of four bioassay test species
Location	Stratified by ecodistrict Restricted by depth and grain size Stratified random
Site number	271 locations 233 reference sites after reviewing criteria for inclusion
Benthic macroinvertebrate descriptors	
Level of identification	Lowest practical level
Number of taxa	162 total taxa: 100 to species level, 62 to genus level, 51 families
Environmental descriptors	
Geographic	5 variables
Limnological	8 variables
Sediment	28 variables

8.1.3 Reference site criteria, site location and number of sites

While optimal reference sites would represent pristine conditions, this objective is unrealistic in the Great Lakes. Initially, minimal impairment was defined as the condition of areas more than 10 km from a point source. The primary source of information on point source discharges is the Canadian Great Lakes Basin Intake–Outfall Atlas (Ontario Ministry of Environment 1990), that describes water intakes and industrial and municipal discharges to the lakes. Avoidance of areas likely to be affected by non-point sources was achieved by using topographic maps to select areas that had minimal agricultural or urban shoreline development. After this first step, fairly large areas were removed from consideration for reference sites.

Since aquatic systems are reflections of the surrounding terrestrial environment a terrestrial land classification system was used to stratify the Great Lakes near shore environment. This classification process described, compared and synthesized data related to the biological and physical characteristics of the land, including parent material, landform, hydrology, vegetation, climate and wildlife. Seventeen ecodistricts intersect the shoreline of the Great Lakes on the Canadian shore, and that level of stratification was used in selecting the reference sites. To identify potential sampling areas, hydrographic charts were used to establish boundaries around areas with either a silt or mud substratum. As the focus of the study was on the near shore environment, a depth criterion was used to limit site location, and thus to set the outer boundary of potential sampling areas. Initially sites were restricted to a maximum of 30 m water depth and within 2 km from the shore. However, fine-grained material was absent within this depth stratum in some geographic areas, so some deeper sites were included. Thus, through four steps, a set of areas was established where reference sites could be located: (i) eliminating areas exposed to stress, (ii) identifying ecodistricts/regions, (iii) delineating areas with fine grained substrate, and (iv) setting a 30 m depth boundary.

The sites were distributed in the areas by assigning them proportionally among the potential sampling areas in the ecodistricts. While there was no *a priori* way of determining the number of reference sites required we aimed for between 250 and 300 reference sites using Wright's (1995) target of 5–10 sites per group and 21 strata (ecodistricts). The average number per stratum was 12, but this ranged from 22 sites in two of the largest strata to as few as 3 in two small strata. Points were located first on the map and then in the lakes. A grid was placed on the potential sampling areas on the map, and a grid square was selected randomly. A site was then located at the center of the selected grid square. This procedure was continued until the assigned number of sites per stratum was occupied. It was sometimes necessary to relocate sites in the field if the substratum turned out to be inappropriate for sampling. Samples were taken from a total of 271 different site locations.

8.1.4 Descriptors of the benthic community and its environment

The fundamental dataset included 162 benthic macroinvertebrate taxa, identified to the lowest practical level. The majority of taxa were not abundant and only 16 taxa contributed more than 1% each of the total number collected. The 10 most abundant taxa comprised more than 70% of all the organisms found. The most diverse groups of organisms identified were the Chironomidae (midge larvae) with 44 genera, the Oligochaeta (worms) with 40 species identified (19 Tubificidae, 18 Naididae and 3 Lumbriculidae) and the Mollusca (snails and clams) with a total of 38 species identified (20 Gastropoda, 18 Bivalvia). Over 40 descriptors of the geographical, limnological, and sediment environment of each site were also recorded.

8.2 Fraser River Biomonitoring Program

8.2.1 Objectives

The river and its catchment occupy a special historic, political and socio-economic niche in British Columbia. The basin's present economy is still dominated by resource extraction, but now the highly urbanized population increasingly depends on the river to process its liquid wastes and on the basin's rivers, lakes and wetlands for recreation. Because of the Fraser River's importance to the province, the Federal Government decided, in 1991, to assist citizens, industries and all levels of government in the development of new ecosystem management practices that would ensure the future sustainability of the basin's ecosystem and economy. All activities were coordinated through the Fraser River Action Plan (FRAP), jointly implemented by Environment Canada and the Department of Fisheries and Oceans between 1991 and 1997.

Developing ecosystem-based management practices requires quantitative measures of ecosystem health. These measures can be used to identify components under stress and to evaluate programs aimed at controlling the stresses. An Environmental Quality Program was designed to assess the basin's present aquatic ecosystem health and the stresses affecting it, and to develop indicators of stress for subsequent management. The program was focused on pollution stress as there were concerns that this stress, already evident in the basin, would lead to a rapid decline in the ecosystem's biological productivity and diversity. There was also a need to establish baseline conditions against which to track the performance of pollution abatement strategies implemented or recommended by the Pollution Abatement Program of FRAP (Environment Canada 1998). A number of the objectives of the program were met by developing a Fraser River biomonitoring program based on benthic assemblages, which was based on the Reference Condition approach.

The decision to use the Reference-Condition approach required the development of a large database of reference sites with information collected on both the assemblage of invertebrates present at each site and on a matching set of habitat descriptors. From this reference database, empirical models could be constructed for predicting the assemblage of invertebrates expected to occur at a site with particular habitat attributes. The reliable prediction of the type of assemblage to be expected at a site could then be used as a baseline against which to assess the status of a site that is suspected of being exposed to either chemical or physical stress.

The overall objectives of the FRAP Environmental Quality Program were to answer the following:

- What is the relative level of stress from contaminants as reflected in concentrations and biological responses at the species, population and community levels?
- What contaminants or classes of contaminants are responsible for the stress?
- Are these contaminants exceeding available guidelines, criteria or objectives for the protection of aquatic life and other uses and are these adequately protected? If a guideline is not available for a contaminant should one be developed?
- Are present pollution abatement programs addressing the contaminants responsible for the stress in the system, and has the ecosystem responded positively to recently implemented abatement programs?
- On what sub-basins or ecosystem components should ambient environmental assessment focus in the next decade?
- Are there new indicator species, biotic communities or other biota descriptors that would improve the assessment of the level of contamination and its impact on aquatic ecosystem health?
- Are there better ways to evaluate the impacts of complex effluents in large rivers than the upstream–downstream approach?

Table 3-4 Study design stage decisions for the Fraser River case study.

Study Design stage	Decision
Spatial extent	Fraser River catchment (250,000 km^2) streams (1st to 7th order on 1:250,000 map
Spatial ggrain	Longitudinal 6 x bank-full width of stream
Temporal extent	3 years, autumn
Temporal grain	1 hour site visit
Criterion for Reference	In the drainage basin, minimal or no logging, mining, agriculture, flow modification or urbanization
Criteria for rejection as Reference	Evidence of any of :channelization, bank alteration, modification of the riparian zone, or upstream disturbance
Location	Stratification based on ecoregion and stream order, then random assignment
Site number	222 sites, 3 removed because of a unique fauna 219 sites in the fundamental dataset
Community data	
Level of identification	Lowest practical level
Number of taxa	323 total taxa: 229 to genus level, 74 families
Environmental data	
Landscape	5 variables
Site/reach	10 variables
Channel/substrate	16 variables
Water-column	9 variables

8.2.2 Spatial and temporal extent and grain-size

The objectives of the study defined the Fraser River catchment as the study area. A "site" was considered to be a stream reach with a longitudinal distance approximately six times its width (Newbury 1984). Only erosional (riffle/run) reaches were sampled because they were the dominant habitat type. Sampling of other habitat types would have taken more time, and multiple-habitat sampling often provides redundant data (Parsons and Norris 1996). The large number of reference sites required a sampling program that was staged over three years: approximately 50 sites were sampled in 1994 and approximately 100 sites in each of 1995 and 1996. The 1994 sites were sampled by one four-person team, whereas the 1995 and 1996 sites were sampled by two four-person teams. Surveys were completed by helicopter within two weeks in autumn to maximize accessibility to streams during the low-flow period; most sites were inaccessible by any other means.

8.2.3 Reference site criteria, site location and number of sites

Actual site selection involved two stages. First, a series of workshops with provincial experts identified non-, moderately, and heavily disturbed subcatchments, and guided the choice of subcatchments that were distributed through as many ecoregions as possible. As a result, 22 of 47 sub-catchments were identified as potential reference areas, unimpacted by point sources, logging, mining or agriculture (see Table 1.3 in Rosenberg et al. 1998). Of the 12 ecoregions contained within the Fraser catchment only one, the Lower Mainland, was not represented in the 22 potential reference subcatchments.

Reference subcatchments suggested for sampling were then outlined on 1:250,000 maps and stream orders were identified using the Strahler (1964) method. Two stratification methods were used to encompass the range of conditions within the study area: (1) ecoregion, which ensured that climatic and landscape conditions were included in the reference sites selected, and was based on previous studies (B.C. Ministry of Environment, Lands and Parks and Environment Canada 1991); and (2) stream order, which ensured that a range of hydraulic conditions was included in the reference sites selected, and was based on Strahler (1964). Two or three sites were randomly located on each stream order within a sub-catchment.

The second stage of site selection was field verification, which involved either land-based or aerial (via helicopter) surveys of the area surrounding the site to confirm that the sites were not disturbed. Occasionally, site locations were moved during the field operation because the original ones were inaccessible or because the identified streams were dry; however, the distribution of sites among stream orders was maintained.

Additionally, a small number of disturbed catchments was included to verify performance of the reference-site model during its development, and 28 sites along the Fraser River were selected at approximately 50-km intervals to provide data for analysis of environmental health in the Fraser main stem. Five of these sites were sampled in 1994, and 14 were sampled in each of 1995 (even-numbered sites) and 1996 (odd-numbered sites).

A total of 222 reference sites was sampled over the three-year study period, together with 48 test sites, 21 of which were related to specific human activities (agriculture, logging, and mining). Three of the 222 reference sites were excluded from further analyses because they originated in a moss-covered habitat and had a unique assemblage of organisms. Therefore, a final set of 219 reference sites was used in the data analyses.

8.2.4 Descriptors of the benthic community and its environment

The 219-site reference database included 323 taxa, of which 136 were dipterans. Of the 323 taxa, 229 could only be identified to the genus level. Most genera were not abundant and only 16 genera contributed more than 2% to the total number of organisms collected. There were 12 common genera (occurrence at >50% of sites), which represented 47% of all the organisms found. The most diverse group of organisms identified was the Chironomidae (midge flies) with 68 genera. The Limnephilidae and Perlodidae had eight genera, and five other families had representatives of seven genera. However, 34 of 74 families were represented by a single genus. The most common and abundant taxon, occurring at more than 80% of the sites, was the mayfly, *Baetis* spp.

Forty environmental variables were measured at each site. These variables were selected based on consideration of relevant literature (see Rosenberg et al. 1998) and represented four different categories (landscape variables, site/reach variables, channel/substrate variables and water-column variables).

8.3 Spring and Autumn Assessment of Streams in the Australian Capital Territory

8.3.1 Objectives

In Australia there has been increased pressure on water managers and the government to maintain ecological values (Norris and Norris, 1995). The Prime Minister's 1992 statement on the Environment allocated $10 million to studies that became the National River Health Program (NRHP) for assessing the ecological condition of Australia's rivers. Central to the NRHP was the development of predictive models, similar to the British RIVPACS models (Wright 1995). The NRHP involves the major environment agency in each State and Territory and is administered by Environment Australia and Land and Water Australia.

The main objectives of the NRHP were to develop and enhance techniques to monitor and assess rivers; to establish a national approach to monitoring and assessing rivers; and to conduct the first national assessment of river health (FNARH). Specific aims of the national program were to develop standardized methods for field sampling, laboratory procedures and data analysis for assessing rivers, using macroinvertebrates and predictive models based on methods proven in the United Kingdom (Wright et al. 1984, Moss et al. 1987). One component of the NRHP was a two-season study of streams in the Australian Capital Territory.

Table 3-5 Study design stage decisions for the Australian Capital Territory case study.

Study Design stage	Decision
Spatial extent	Australian Capital Territory Streams and adjacent streams in the upper Murrumbidgee catchment (1st to 5th order on 1:50,000 map)
Spatial ggrain	Longitudinal 10 x bank-full width of stream
Temporal extent	2 years, autumn, spring
Temporal grain	1 hour site visit
Criterion for Reference	>50 m upstream and >300 m downstream of weir, dam, ford, waterfall >5 m high, significant discharge or confluence, livestock watering area, areas subject to flow diversion, channelization, dredging or weed removal
Criteria for rejection as Reference	Gross pollution from single point sources (e.g., sewage treatment plants) Diffuse but intense pollution (e.g., storm water, intensive agriculture)
Location	Stream types, geology, vegetation types and land uses
Site number	93
Community data	
Level of identification	Mostly family Oligochaeta – class Acarina – order Chironomidae – sub-family
Number of taxa	Autumn – 64 total, 41 common Spring – 61 total, 35 common
Environmental data	
Landscape	6 variables
Site/reach	19 variables
Channel/substrate	6 variables
Water-column	3 variables

8.3.2 Spatial and temporal extent and grain-size

The Australian Capital Territory component of the NRHP involved the sampling and assessment of rivers within the Murrumbidgee River catchment above Burrinjuck Dam wall, including the catchments within the Australian Capital Territory, and the Yass River, Molonglo River and Queanbeyan River catchments. In all, there were four sampling rounds for the Australian Capital Territory part of the program between autumn 1994 and spring 1995.

8.3.3 Reference site criteria, site location and number of sites

Initial reference site screening was based on an extensive pilot study in which almost 100 sites were sampled throughout the Australian Capital Territory and surrounding area. The region was divided into major units based on altitude, slope, geology, vegetation type and cover and land uses, catchment area and stream order.

The pilot study sites were allocated proportionally to each of these categories. They therefore included sites representing characteristics that would ultimately be deemed reference and test site conditions. Habitat characteristics were recorded at these sites using the USEPA methods (Plafkin et al. 1989) and the habitat condition of sites was assessed on the basis of whether a site was likely to be able to support a healthy community of invertebrates. This procedure followed the original USEPA protocols that required the habitat to be capable of supporting a healthy community before further testing was carried out. Following this initial screening, the reference site selection process provided by Davies (1994) and explained in Section 3.3 (Figure 3-4)[check new numbers] was implemented. It specifies that reference sites should cover a wide range of physical and chemical conditions encompassing all types of river discharge, size and altitude within the study area. Reference sites were chosen on the basis that they were minimally disturbed, with lowland rivers expected to be more affected by a range of land use practices. Reference sites were required to be representative of the reach (10 X bank-full width), not subject to significant flow regulation, and accessible and safe during sampling.

Additionally, reference sites were not chosen in the area 50 m upstream to 300 m downstream of a weir, dam, ford, waterfall greater than 5 m tall, significant discharge or confluence, livestock watering area, areas subject to flow diversion, channelization, dredging or weed removal (Davies 1994). Grazing was not considered a land use detrimental to water quality, unless stock had access to the river banks, or the riparian zone was degraded.

Test sites were selected based on potential and known impacts and also incorporated sites used within existing biological monitoring programs. Potential and known impacts ranged from agricultural runoff, rural degradation, urban runoff and discharge of treated sewage effluent, to metals, dams, salinity and sand mining. Characteristics of the reference and test sites for the two sampling areas, the Australian Capital Territory and surrounding region and the Upper Murrumbidgee River Catchment (excluding the Australian Capital Territory and surrounding region) are listed in Table 3-1a, b[check new number is correct].

After the first sampling rounds were complete the data from each site were reviewed (see Chapter 4). Any sites that were suspect because of few taxa, or a numerical dominance of Oligochaeta and/or Diptera were re-considered. First, the field notes on data collection were checked to see if any factors might have affected the collection (e.g., adverse weather or flow conditions, an unnoticed hole in a net), or if there were features of the site not accounted for such as stock access, or an upstream point source. After this screening the final set of sites was agreed upon, in consultation with the local agency staff.

The Australian Capital Territory sampling covered 100 reference and 20 test sites. These latter sites were ones known to have major impacts and were sampled to make a first coarse assessment of the models derived from the reference sites. After reviewing the initial hundred reference sites, seven were rejected as having previously unknown impacts, or some problem with sampling — usually lack of water because of prevailing drought conditions.

8.3.4 Descriptors of the benthic community and its environment

There was a pool of 87 taxa (mostly families) of which over 60 occurred in either spring or autumn. The taxa that occurred at fewer than 10 sites were eliminated when building the model, leaving 35–40 commonly occurring taxa.

Eighty two environmental variables were recorded for each site but many of these were deemed likely to be affected by human activities in the study region and were omitted from the model building. This left a final list of 34 variables that were used for constructing the model (Chapter 4).

9. POINTS TO REMEMBER

- The objectives of the study need to be defined with all the interested parties contributing.
- The extent and grain of the study need to be determined, remembering that reference sites may need to be located outside the immediate area of interest and that there may be seasonal and inter-annual variation to account for.
- The grain size, or the criteria for choosing the characteristics that represent a site, need to be determined up-front.
- The site, rather than points within it, is the grain size in the study design. The site is the scale of sampling and replication.
- A wide range of environmental descriptors should be chosen at a variety of scales. Many large-scale descriptors are important for predictions.

— Criteria for a reference site need to be decided up-front and applied rigorously. The steps will probably include a review process and will recognize that acceptable sites may be subject to some change caused by humans.
— Sufficient reference sites should be chosen to represent the range of habitat types and communities likely to be encountered in test site sampling.
— Specific biological descriptors need to be chosen up-front. These may be represented differently (indices, metrics, functional feeding groups, abundance, composition) depending on how sampling is done.
— Sampling effort must be standardized and equivalent at both reference and test sites.
— As far as possible, sampling circumstances at reference sites will dictate sampling at test sites, e.g., same season.
— An efficient data storage system needs to be designed at the start of the study and data quality control and assurance systems should be implemented from the outset.

Chapter 4

VARIATION IN THE REFERENCE CONDITION

Description and modeling

1. WHY CHARACTERIZING REFERENCE SITE VARIATION IS IMPORTANT

In the past, variability in the assessment of freshwater ecosystems has commonly been seen as a problem to be addressed in study design (Chapter 3) by restricting sampling spatially and temporally (e.g., upstream and downstream of an effluent in Autumn), and to particular habitats (e.g., profundal zones in lakes). In contrast, when using the Reference Condition Approach, we want to be sure that we adequately sample the natural variability among the freshwater ecosystems present in the region of interest (e.g., Simpson and Norris 2000). Then, an ecologically relevant comparison can be made between a particular test site (or set of sites exposed to some common stressor) and the appropriate Reference Condition. Variation in the Reference Condition, rather than being a problem, is a characteristic to be "captured" for the region being studied. We must measured and characterize it before we can assess a test site.

Within-site variation in benthic macroinvertebrate populations and communities has been recognized for some time in both pure and applied ecological research, and has been used to determine the sampling effort necessary to characterize a site (Needham and Usinger 1956; Downing 1979; Resh 1979; Allan 1984; Morin 1985; Norris and Georges 1986; Canton and Chadwick 1988; Norris et al. 1992, Resh and McElravy 1993, Norris and Georges 1993). Less effort has been made to characterize larger spatial and temporal scale variation in the biota and their environment *among* sites. In the Reference Condition Approach the site is the scale of replication, and, as we showed in Chapter 3, variation among sites is much greater than variation within sites. We believe that the first step of any Reference Condition Approach assessment should be a careful inspection to see how the Reference Condition varies with respect to both the biota and the environments of reference sites. In other words, we want a clear notion of how and by how much the Reference Condition varies.

2. HOW TO DESCRIBE VARIATION AMONG REFERENCE SITES

2.1 Data examination

To show the basic properties of the biota or their environment, we can use a simple histogram or box plot, together with descriptive statistics such as the mean or median and standard deviation or quartiles. Pairs of biological descriptors (such as Taxon Richness and total abundance) or environmental descriptors (such as conductivity and alkalinity) can be informatively plotted as bivariate scatter plots, which simultaneously portray the central tendency and variability of each descriptor, as well as their covariation. To describe the variation and covariation of several biotic or environmental descriptors simultaneously, we can use a SPLOM plot (scatter plot matrix; Quinn and Keough 2002) or a variety of ordination techniques including, perhaps most commonly, principal component analysis or non-metric multidimensional scaling (see Section 2.2). At this stage we do not statistically associate environmental and biological measures. We are simply developing a familiarity with the properties of the Reference Condition.

By examining the basic properties of the data, we can determine the context and boundaries of a study. Information such as the range of expected conductivities or pH, the possible number of taxa and their seasonal and inter-annual stability will help us understand the system and interpret subsequent analyses.

2.2 Classification and ordination of the range of Reference Conditions in the biota

Two methods are principally used to explore the variability among sites, and for distinguishing the range of possible Reference Conditions: ordination and classification.

Ordination is a useful way of graphically representing variation in the biota or environments of a range of sites in two or three dimensions, and it has been used extensively in studies of benthic macroinvertebrates (see Resh and McElravy 1993). There are many approaches to ordination that are introduced in the readable text by Manly (1986), and described in some detail by Gauch (1982), Pielou (1984), and Legendre and Legendre (1998). A useful summary of the attributes of various ordination methods is provided by Norris and Georges (1993).

Classification can be applied *a priori*, with ecoregions for example (Omernik 1995), or *a posteriori*, using one of many possible clustering procedures. In six of seven examples where *a priori* and *a posteriori* classifications were tested, more variability in reference sites was accounted for by an *a posteriori* classification using the biota sampled at reference sites to form groups (Hawkins et al. 2001). *A posteriori* classification of the biota from sites considered to represent Reference Conditions has been used widely in Great Britain (Moss et al. 1987), Australia (Simpson and Norris 2002), Canada (Reynoldson et al. 1995, 1997) and the USA (Hawkins et al. 2000a), as a precursor to developing predictive models. The Reference Condition, as we apply it, uses rigorous criteria for the selection of sites (Chapter 3). Thus, once we come to describing variability in the Reference Condition in this section, we already have the data in hand and we apply *a posteriori* classifications before subsequent analyses.

There are several decisions to be made before applying classification or ordination to sites, based on the biota. These include:

a) inclusion or exclusion of rare taxa from the analysis (Norris et al. 1982, Marchant et al. 1984),
b) measures used to represent each taxon in calculating the distance matrix (e.g. biomass, relative abundance, presence/absence),
c) whether to apply data transformation, which will have the effect of up- and down-weighting rare or abundant taxa, and
d) which distance measure to use (e.g., Bray & Curtis, Jaccard's),
e) the number of clusters or groups to be determined in a classification, or the number of axes to use in an ordination (Faith et al. 1987, Belbin and McDonald 1993, Norris and Georges 1993, Legendre and Legendre 1998).

(a) Inclusion of rare taxa can be important in univariate and graphical methods of analysis, because omitting them will affect the analysis outcome (if using Taxon Richness, for example). There are two circumstances in multivariate analysis where it is desirable to eliminate rare taxa. First, when using principal component analysis (PCA) ordinations (not multi-dimensional scaling (MDS)) the number of taxa sometimes needs to be reduced to avoid problems in computing eigenvalues. Second, while it is not difficult to construct dendrograms or MDS ordinations using all taxa, the rare taxa, whose occurrence at a station largely results from chance, tend to add noise and thus obscure the patterns in the data (Clarke and Warwick 1994).

The way in which taxa are eliminated requires careful consideration. A common method is to remove taxa contributing to less than 1 or 2% of the total abundance, but this criterion can be problematic where the total abundance between sites is very variable. Also, sites with very low numbers may be characterized by certain taxa. A better approach is to retain all taxa that account for >*p*% (e.g., 1–4%) of the total abundance in any one sample. In agreement with Legendre and Legendre (1998), we consider that joint absences do not carry much ecological information. Also, they tend to contribute noise, or variability, to the data that it is very difficult to account for and will often weaken analyses. We emphasize that, in using the Reference Condition Approach for assessment, we are making comparisons. The important consideration is the amount of information needed for a robust comparison and to account for and use variability in the Reference Condition — the subject of this section. Rare taxa, by their nature, are usually scattered through the data matrices and provide little useful information for comparisons. This is not denying their importance in other situations, such as conservation and biodiversity studies (e.g., Faith and Norris 1989), but those needs should not be confused with making robust comparisons for assessment.

(b) There are three main options for representing the taxa: absolute abundance (as numbers or biomass per unit area), relative numbers or biomass, or reduction to presence and absence. Absolute numbers are most commonly used where quantitative collections have been made. They provide complete information on the taxon observations. Relative numbers provide a standardized value (for example, percentage of total abundance). Such standardization may be appropriate where samples have been sub-sampled, say for standardized counts. Reduction to presence and absence may be justifiable if samples have been sorted to maximize the taxonomic list without an attempt to count numbers (e.g., Wright et al. 1984), or where the analysis will only be concerned with presence and absence data.

(c) Data transformation is not required to meet the data assumptions in many analyses. However, it can be used to weight the contributions of common and rare taxa. The choice of transformation can affect the conclusions of an analysis and is often more a biological or ecological question than a statistical one. It will in part be determined by the nature of the data (e.g., sub-sampling or qualitative collections).

(d) The first step in classification or ordination is to choose a similarity index to calculate the degree of similarity of any two sites in their biota or environment and thus create a similarity. While numerous similarity measures have been developed, the one most commonly used in ecological studies is the Bray–Curtis index (Faith et al. 1987). Many reviewers have indicated that the Bray–Curtis index (Bray and Curtis 1957) is the most reliable (Bloom 1981, Marchant et al. 1984, Pontasch et al. 1989, Jackson 1993). Faith et al (1991) showed that for detecting effects of uranium and gold mines the Bray–Curtis index was superior to seven other indices. It provided consistently high statistical power because of its low susceptibility to temporal variability at control sites and its sensitive response to disturbance. It is also unaffected by the nature of communities being compared (Bloom 1981) and differences contribute in the same way to the Bray–Curtis index regardless of the species being rare or abundant. The Bray–Curtis index is generally suitable because it takes a maximum value when two samples are identical, it takes the value 0 when samples have no taxa in common, changes in measurement units do not affect its value, its value is unaffected by inclusion or exclusion of a species jointly absent from a pair of samples, inclusion of a third sample makes no difference to similarity between previous samples and it can detect differences between total abundances even when relative abundances are identical.

(e) Determination of the number of groups resulting from a classification is subjective, in that we identify significant gaps in the classification tree or dendrogram that indicate greater similarity within groups than among groups. Applications of the RCA in have tended to use five (Simpson and Norris 2000) or ten (Reynoldson and Wright 2000) reference sites per group as a minimum, depending on other aspects of the analysis. There will be a trade-off between accounting for variation in the reference sites and the number of groups.

3. CASE STUDIES: VARIATION AMONG REFERENCE SITES

To explain the concepts and measures outlined above, we now describe their use in our three case studies, at the North American Great Lakes, the Fraser River and the streams of the Australian Capital Territory.

For each reference site in the case studies, we use Taxon Richness and the Bray–Curtis distance of the community from the average reference community (which we call MCDist) as single descriptors of the biota. These two descriptors indicate how the Reference Condition varies in each study. In Section 4 of this chapter, when constructing predictive models from the case studies, we use only MCDist as an example of a single descriptor of the biota. In our initial graphical analysis for the case studies, we apply several commonly used water quality and habitat descriptors to characterize variation among reference sites in their environments.

For classification and ordination of each of the case study datasets, we first eliminate the rare taxa (see Norris et al. 1982, Marchant et al. 1984, Simpson and Norris 2000); next we take the logarithm of relative abundance of the non-rare taxa; then we calculate the Bray–Curtis distance between all pairs of reference site communities; and finally we apply clustering using UPGMA with $\beta = -0.1$. For ordinations, we use non-metric multidimensional scaling and derive two dimensions. In omitting rare taxa we use occurrence at sites, rather than a proportion of abundance, as the criterion because the sites are the level of replication and the AUSRIVAS method considers only presence and absence data. Thus, we eliminate taxa that occur at <10% of the sites if the dataset has <100 sites, or those that occur at <10 sites if there are >100 sites in the dataset.

3.1 Near shore Environment of North American Great Lakes

As was expected in the North American Great Lakes, with their striking upstream (Lake Superior and Lake Michigan) to downstream (Lake Erie and Lake Ontario) gradient in climate, surficial geology, water chemistry, trophic status and resultant biogeographic region, both the biota (Figure 4-1) and the environment (Figure 4-2) of sites in Reference Condition vary substantially. From 2 to 20 taxa were found at each reference site (richness mean = 7.7), and the Bray–Curtis distance of each site to the median community (MCDist) varies from 0.28 to 0.98 (Bray–Curtis minimum and maximum values are 0 and 1 respectively). There is no significant correlation between these two descriptors of the benthic community: thus a site that is "odd" in its composition (with a high MCDist value) does not necessarily have more or fewer taxa than average. There is also considerable variation within the lakes in both MCDist and Taxon Richness. It is clear that simple measures such as the number of taxa by themselves are unlikely to adequately describe the condition of a site because of the wide range in number of taxa that represent the Reference Condition.

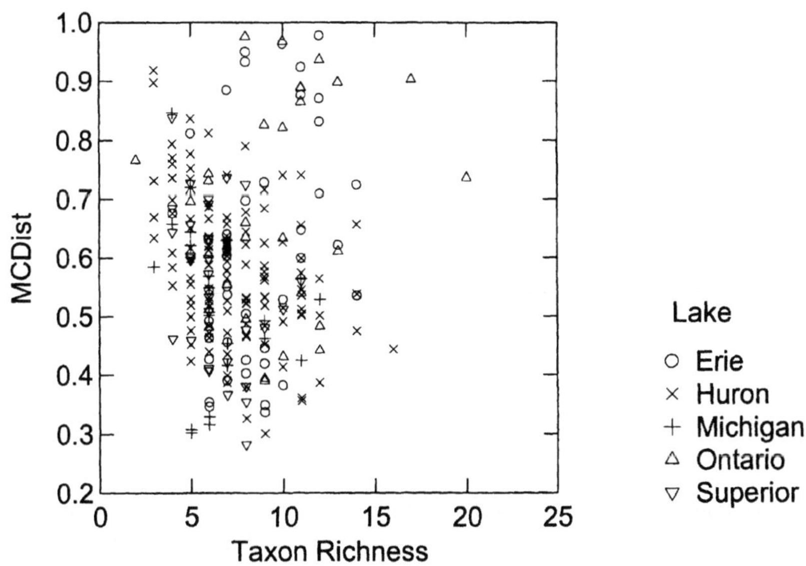

Figure 4-1. Variation in biota among Near shore Great Lakes sites as measured by distance from the median community (MCDist) and number of taxa (family and higher; Taxon Richness) in relation to lake.

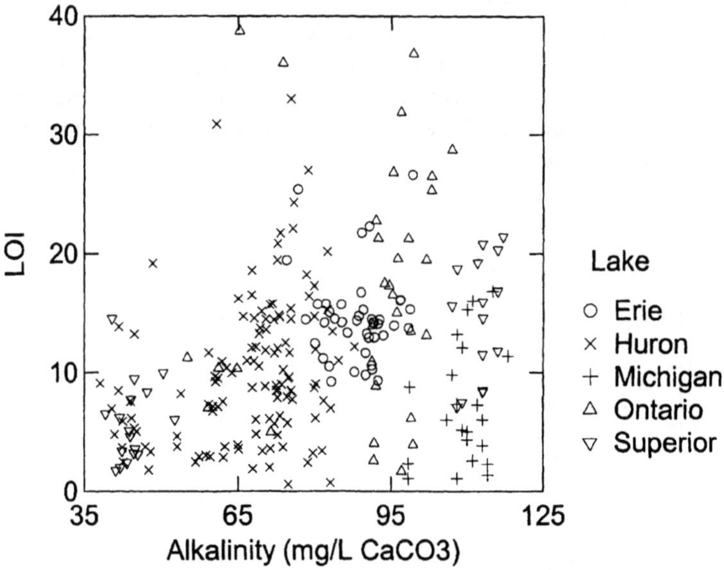

Figure 4-2. Variation in environments among Near shore Great Lakes sites as measured by sediment organic matter (LOI) and alkalinity in relation to lake.

We use alkalinity and organic matter (LOI) in the sediment of the Great Lakes to illustrate variability in environmental variables. These two variables have a weak, positive correlation among reference sites (r = 0.29; Figure 4-2). Along with other environmental variables, they tend to vary among the lakes, indicating the large-scale geographic variation in the drainage basins of these lakes, from the deep, cold, and oligotrophic Lake Superior to the relatively shallow, meso- to eutrophic Lake Erie and mesotrophic (yet deep) Lake Ontario, and reflecting the influences of the Precambrian Canadian Shield on the upper lakes and glacial till on the lower lakes.

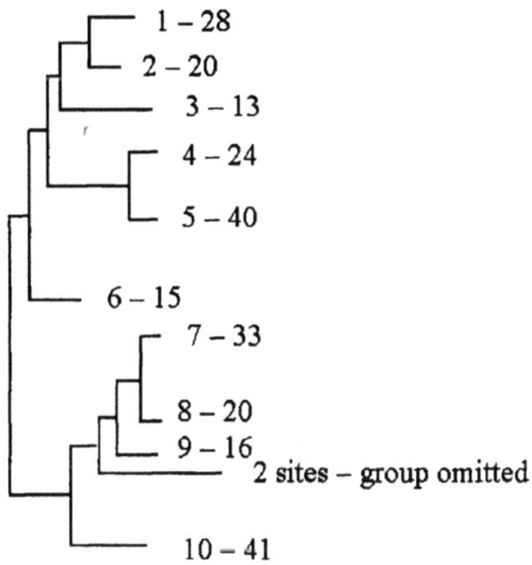

Figure 4-3. Classification dendrogram of Near shore Great Lakes reference sites based on (log-transformed, relative abundance) benthic communities. Each branch of the dendrogram shows Group number – number of sites (e.g. Group 4 had 24 sites).

Classification of the Great Lakes reference sites using the benthic community results in 10 groups of sites with >10 sites each that we use for further analysis (Figure 4-3). To some extent, biota at the reference sites are structured by lake (Table 4-1); most sites from each lake tend to fall in one of three or four of the faunally defined groups, although Lake Huron communities are more heterogeneous, likely due to the variation among the main lake, the North Channel and Georgian Bay. There is at least one Lake Huron site in each of the 10 groups defined by the classification analysis. One group has only two sites and is omitted from further analyses.

Table 4-1. Near shore Great Lakes communities: Faunal group in relation to a reference site's lake.

Group	Superior	Michigan	Huron	Erie/St Clair	Ontario	*Total*
1	0	1	10	17	0	28
2	0	2	14	0	4	20
3	0	0	1	12	0	13
4	0	0	6	12	6	24
5	0	0	26	0	14	40
6	0	1	13	1	0	15
7	8	7	17	0	1	33
8	0	0	15	0	5	20
9	6	3	6	0	1	16
10	18	8	14	0	1	41

Ordination (non-metric MDS, Clarke and Warwick 1994) of the reference communities allows us to visualize in two dimensions the degree of scatter within and among the groups determined from the classification (Figure 4-4). Group 4 communities, which were sampled in Lakes Erie, St Clair, Huron, and Ontario (Table 4-1), are obviously quite heterogeneous in composition. Group 10 communities, which were only found in the deeper, more oligotrophic lakes (Michigan, Superior, Huron, and Ontario), are much less variable.

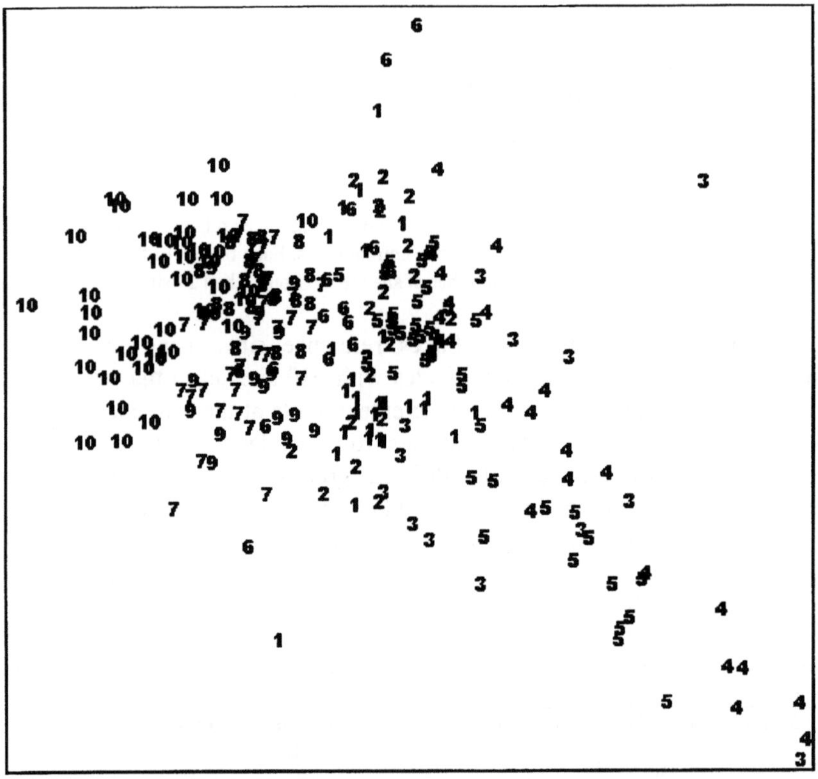

Figure 4-4. Two-dimensional ordination of reference communities from the Near shore Great Lakes study. Plotting points are group numbers determined from the classification (see Figure 4-3).

3.2 Fraser River Biomonitoring Program

Biota from reference sites in the Fraser River Biomonitoring Program have a range similar to that in the Great Lakes (5–25 taxa per site), but a higher average value (mean = 15.3) for Taxon Richness. The range and mean of MCDist (mean = 0.61, range 0.25–0.97) are similar to communities in the Near shore Great Lakes study. As with the Great Lakes biota, there is no discernible correlation between Taxon Richness and MCDist (Figure 4-5). Also, there is no obvious relationship between the value of either variable and stream order.

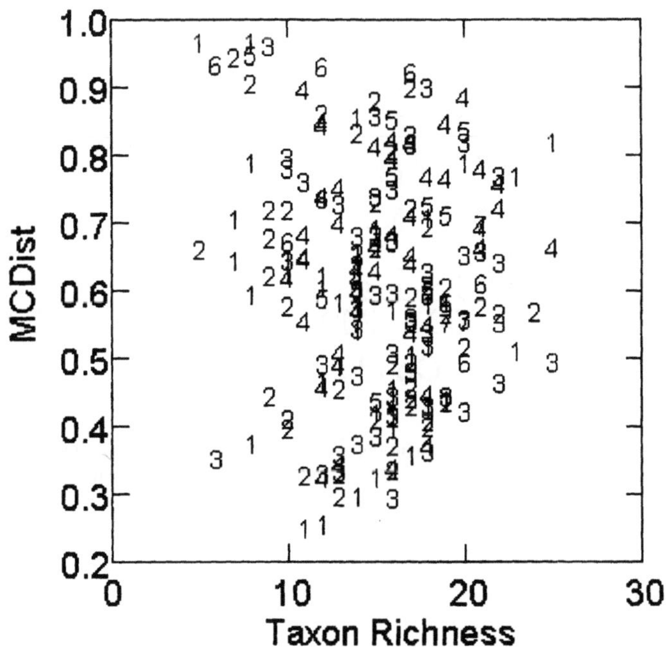

Figure 4-5. Variation in biota among Fraser River Biomonitoring Program reference sites as measured by distance from the median community (MCDist) and number of taxa (family and higher; Taxon Richness) in relation to stream order (point labels).

The environments of Fraser River reference streams also vary widely. Percent gravel at a site, for example, ranges from 0 to 100%, and conductivity varies from a low of 5 μScm^{-1} to over 400 μScm^{-1}. There is a small, negative correlation between these two environmental descriptors (i.e., high conductivity streams have less gravel), but little obvious relationship between stream order and either conductivity or the proportion of gravel substrate (Figure 4-6), perhaps giving some insight into the lack of relationship between the biological descriptors noted above.

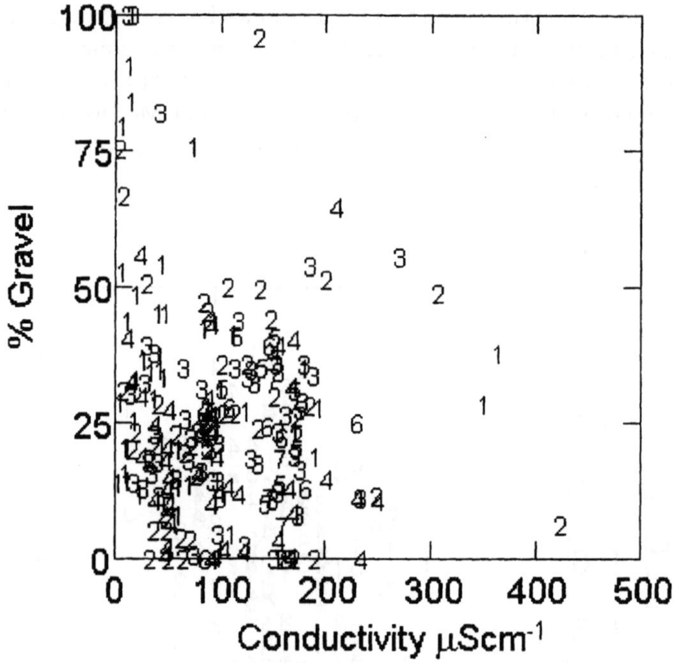

Figure 4-6. Variation in environments among Fraser River Biomonitoring Program reference sites as measured by %gravel in the substrate and conductivity in relation to stream order (point labels).

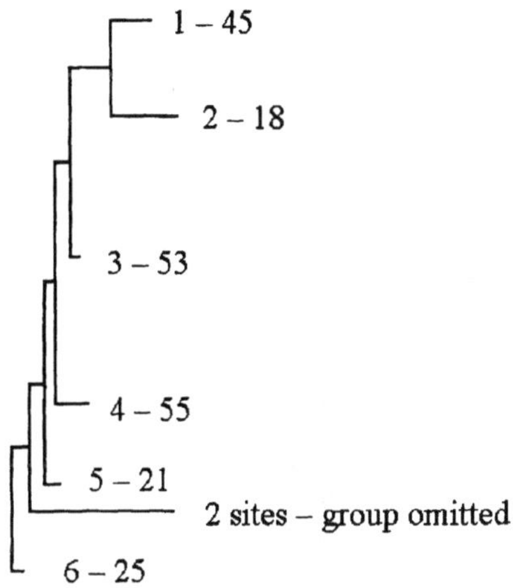

Figure 4-7. Classification dendrogram of Fraser River Biomonitoring Project reference sites based on (log-transformed, relative abundance) benthic communities. Each branch of the dendrogram shows Group number – number of sites (e.g. Group 3 had 53 sites)

Six groups of reference sites, based on relative homogeneity of biota, can be defined from the Fraser River Catchment (Figure 4-7). A seventh group, representing a mossy habitat, had only two sites and is omitted from subsequent analyses. The groups tend to represent the continuum from low-order to high-order streams, distinct at either end but each containing sites from several stream orders. Ordination of the reference communities shows the heterogeneous composition of the Group 6 communities, particularly as contrasted to the small amount of variation among the Group 1 communities (Figure 4-8). Thus, the representation of the biota in an ordination provides an insight into factors such as stream order, that may structure the data when composition is considered, but not simple measures such as taxonomic richness, as noted above.

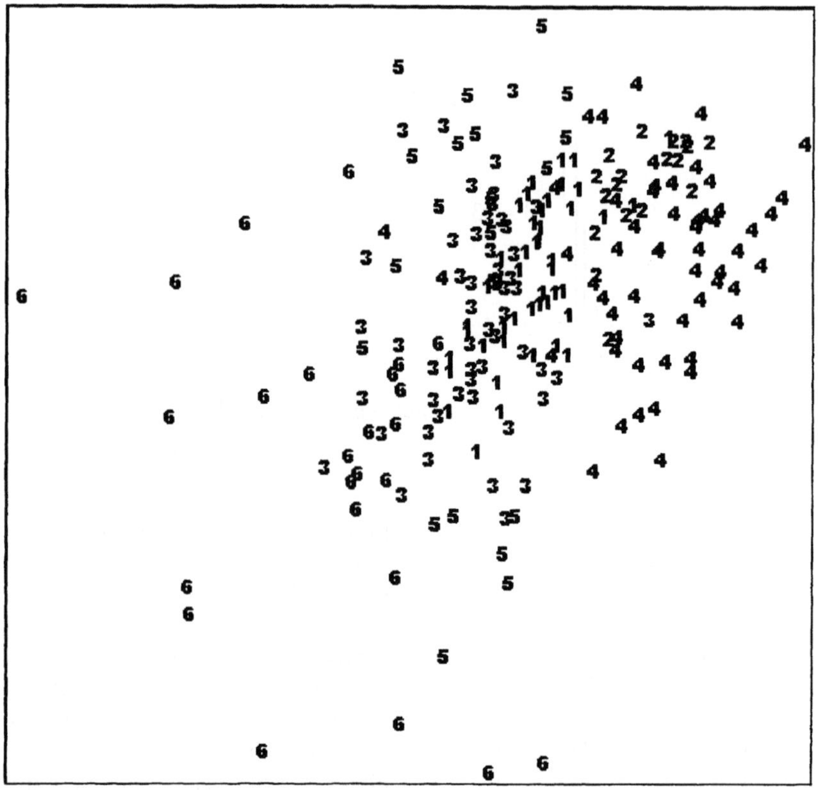

Figure 4-8. Two-dimensional ordination of reference communities from the Fraser River Basin study. Plotting points are group numbers determined from the classification (see Figure 4-7)

3.3 Spring and Autumn Assessment of Streams in the Australian Capital Territory

The effects of season on variability in the Reference Condition are demonstrated in the Australian Capital Territory study of streams and the analyses presented here emphasize these seasonal comparisons. There are surprisingly small correlations between water chemistry variables measured in the Autumn and the Spring (Figures 4-9, 4-10). Many streams are substantially warmer by Autumn than expected from their Spring temperatures (Figure 4-9) indicating the variable Spring weather and its patchy effect on warming of surface-draining streams. Conductivity in the

Autumn and Spring is more highly correlated than temperature (r = 0.82 for Spring and Autumn conductivity *versus* r = 0.42 for Spring and Autumn temperature). With a couple of notable exceptions (Sites R123, R126), conductivity in the Spring is a reasonable measure of what it would be in Autumn (Figure 4-10). This relative constancy tells that conductivity is likely to be related to catchment features and if it is related to descriptors of the biota it is likely to be a more stable predictor than temperature.

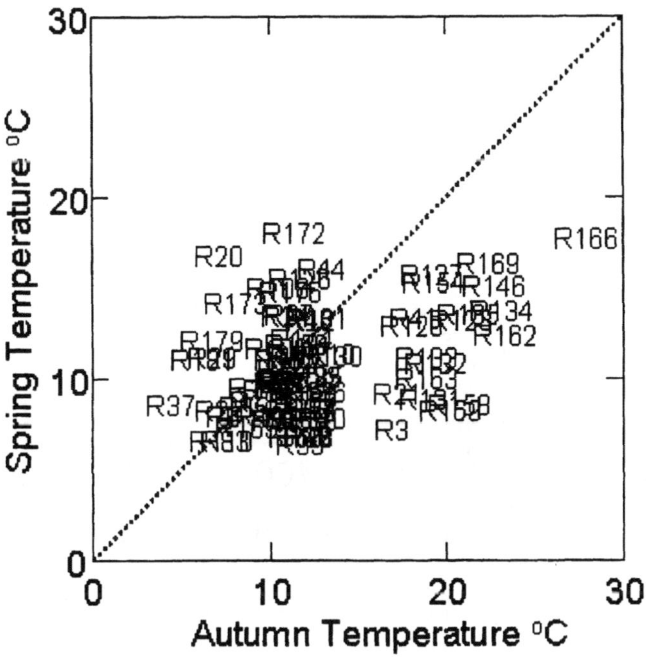

Figure 4-9. Water temperature differences between Spring and Autumn in Australian Capital Territory reference streams.

Taxon (family) richness is usually greater at a given site in the Autumn than during the Spring (Figure 4-11). It is also more variable among the reference sites in Autumn than in Spring; Taxon Richness varies from 7 or 8 to almost 30 taxa at reference sites in the Autumn, and only between 10 and 25 taxa in the Spring. There is only a modest linear correlation between Taxon Richness in Autumn and Spring (r = 0.45), but there is somewhat more variability among the communities in Autumn than in Spring, and little correlation between a community's distance from the median or typical community (MCDist) in Autumn *versus* Spring (Figure 4-12). These results indicate that we may expect assessments based on sampling in the different seasons to differ depending on how well the variability in each season can be accounted for.

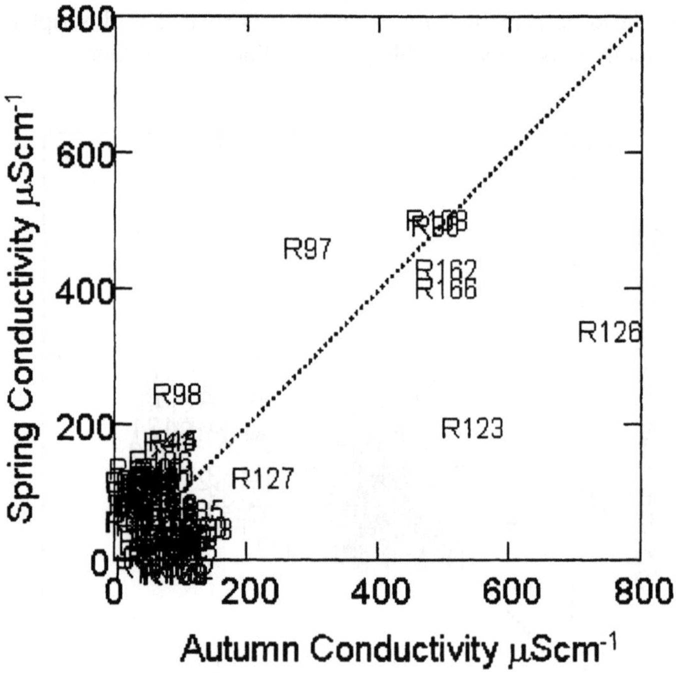

Figure 4-10. Conductivity differences between Spring and Autumn in Australian Capital Territory reference streams.

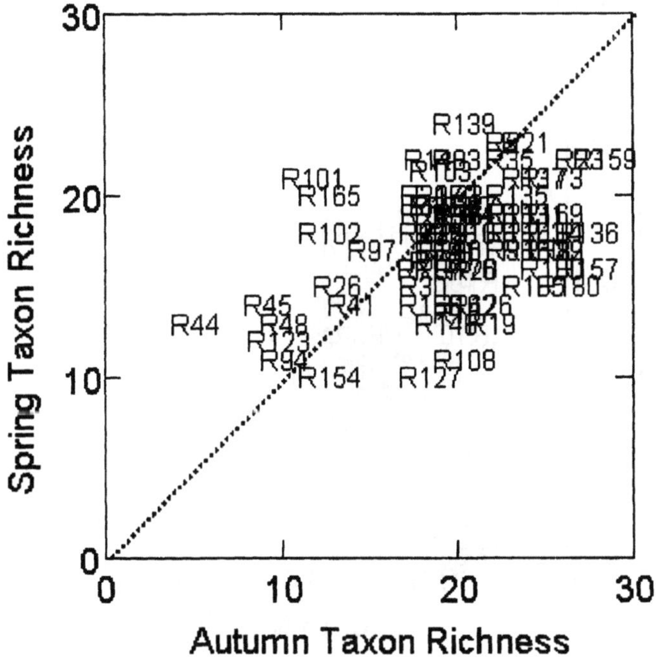

Figure 4-11. Benthic macroinvertebrate Taxon Richness differences between Spring and Autumn in Australian Capital Territory reference streams.

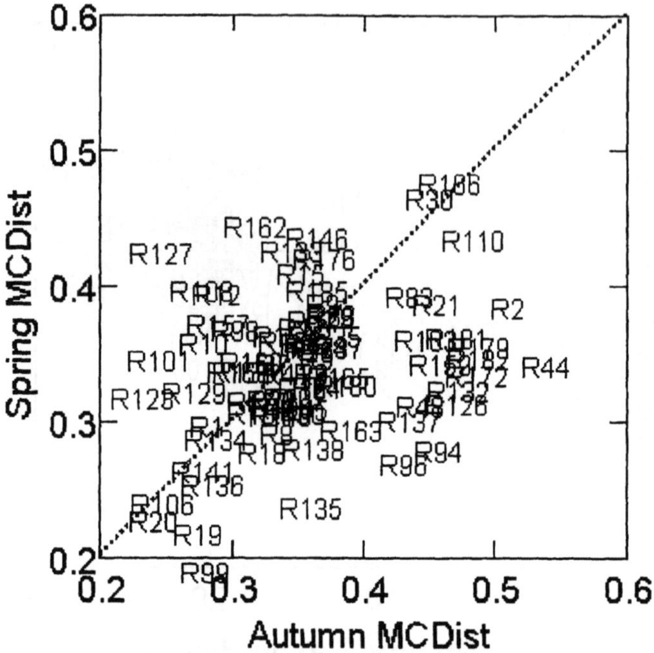

Figure 4-12. Distance from median community (MCDist) of benthic macroinvertebrate communities in Spring *versus* Autumn in Australian Capital Territory reference streams.

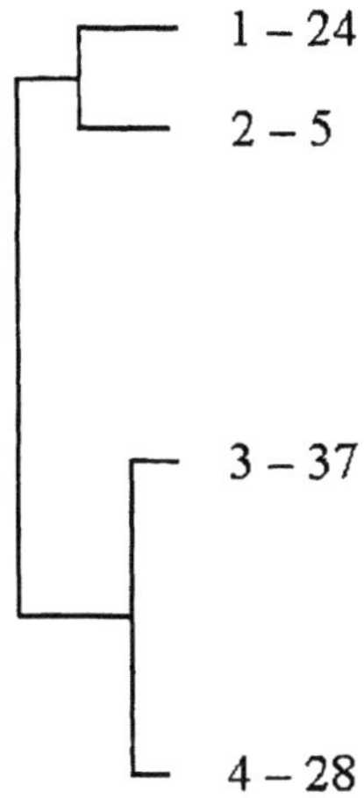

Figure 4-13. Faunal groups in classification of Australian Capital Territory reference sites based on (log-transformed, relative abundance) benthic communities collected in Autumn. . Each branch of the dendrogram shows Group number – number of sites (e.g. Group 2 had 5 sites)

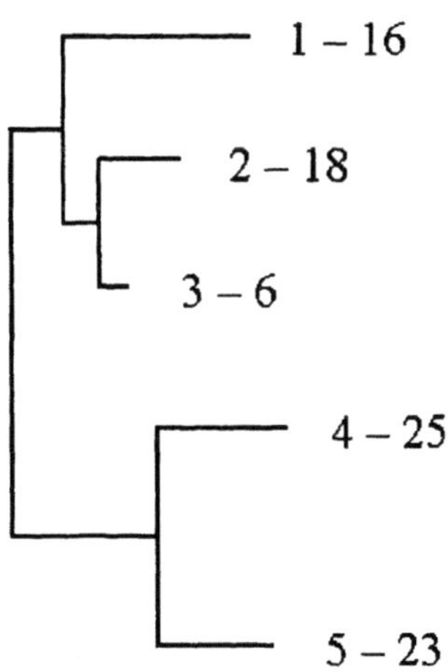

Figure 4-14. Faunal groups in classification of Australian Capital Territory reference sites based on (log-transformed, relative abundance) benthic communities collected in Spring. . Each branch of the dendrogram shows Group number – number of sites (e.g. Group 4 had 25 sites)

The faunal groups determined for reference sites in the Australian Capital Territory study (Figures 4-13, 4-14) are quite consistent in the two seasons (Table 4-2). Most members of Group 1 from the Autumn sample fall into Group 1 of the Spring classification, whereas the two Autumn Group 2 communities fall into Spring Group 5, and Autumn Group 3 members are mainly in Spring Groups 4 and 5. Autumn Group 4 communities are more scattered among the Spring groups..

Table 4-2. Consistency of classification based on Spring (rows) and Autumn (columns) benthic communities. Sites not sampled in both seasons are not included in the table.

Group	1	2	3	4	Total
1	12	0	1	3	16
2	2	0	3	8	13
3	1	0	0	4	5
4	7	0	12	5	24
5	0	2	17	2	21
Total	22	2	33	22	79

The contrasting ordinations of Autumn (Figure 4-15) and Spring (Figure 4-16) communities from the Australian Capital Territory illustrate differences in homogeneity of the classification groups between the two seasons. Although there is substantial variability within groups in both seasons, they are better defined in the Autumn than the Spring. All else being equal, it is easier to detect a test site's deviation from its predicted reference group in the Autumn than the Spring.

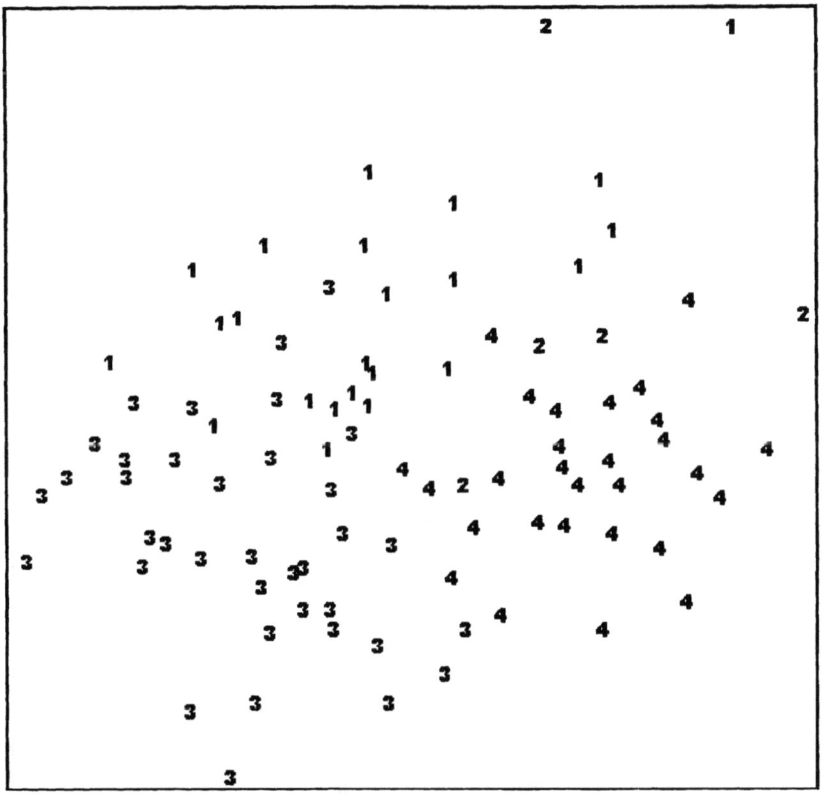

Figure 4-15. Two-dimensional ordination of reference communities from the Autumn sampling in the Australian Capital Territory study. Plotting points are group numbers determined from the classification (see Figure 4-14)

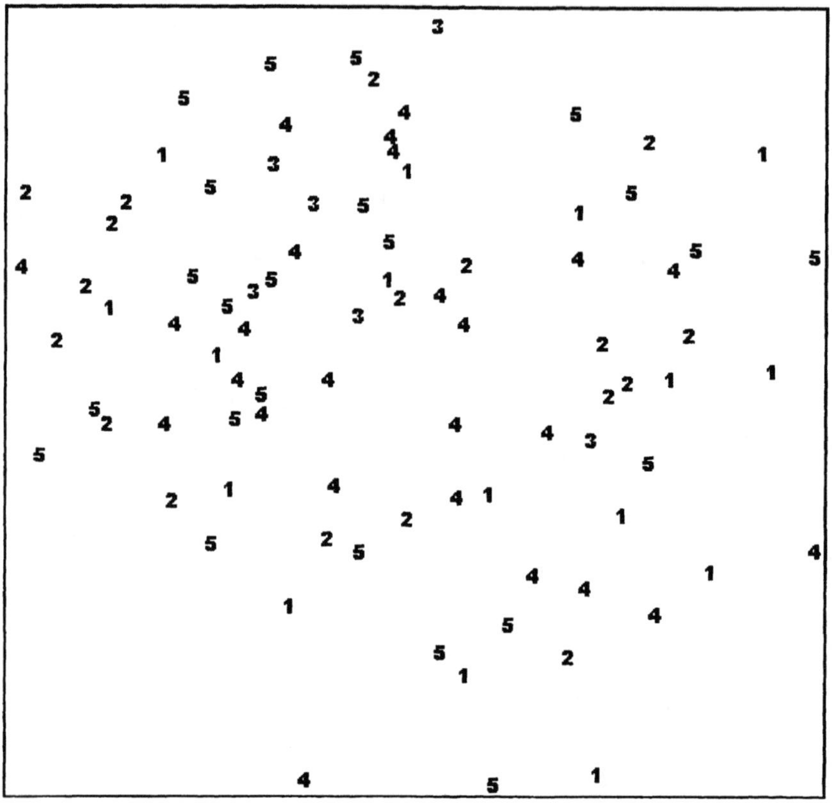

Figure 4-16. Two-dimensional ordination of reference communities from the Spring sampling in the Australian Capital Territory study. Plotting points are group numbers determined from the classification (see Figure 4-15).

4. WHY IT IS IMPORTANT TO MODEL AND EXPLAINE VARIATION AMONG REFERENCE SITES

Once we have described the variation among reference sites in both their environments and their biota, we can establish relationships between the biota and the environment, in an effort to explain as much of the observed variation in the biota as possible. Most ordinations or classifications, and their subsequent use, are based on the assumption that environmental features representing the habitat have a deterministic relationship with the organisms

(Southwood 1977, Marchant et al. 1984, Lake and Barmuta 1986, Townsend and Hildrew 1994). Thus, the properties of any given habitat within an aquatic ecosystem are assumed to determine the types of macroinvertebrate assemblages found there. Once detected, deterministic relationships enable us to predict the fauna that should be found at other, non-reference (test) sites with similar environmental characteristics. This is sometimes referred to as "site matching"; the biota expected at a test site is determined by matching the test site with a real or hypothetical reference site (or set of reference sites) with similar environmental characteristics.

A simple example (Figure 4-17) illustrates the value of modeling variation in the reference biota before assessing a test site. If reference streams have been sampled and a strong relationship between the biota (as measured by Taxon Richness) and the environment (as measured by stream flow) is measured, we can refine our expectation of Taxon Richness at a test site if it is in Reference Condition. Without the relationship and the model built from it, our expectation at a test site would just be the mean Taxon Richness of the reference communities. Such a prediction would have considerable variability associated with it, as indicated by the distribution curve shown on the y-axis of Figure 4-14. Using the model relating stream flow to Taxon Richness, we would use the measured stream flow at a test site to refine our prediction of Taxon Richness. The width of the distribution curve measuring variability in the model (the dashed curve) relative to the original distribution of Taxon Richness among reference sites is a measure of the strength of the relationship between the biota and the environment and the degree to which we can refine our expectation of the biota. The circled site in Figure 4-17 has a Taxon Richness well above the mean in the reference distribution but not unexpectedly high, given the high stream flow at the site.

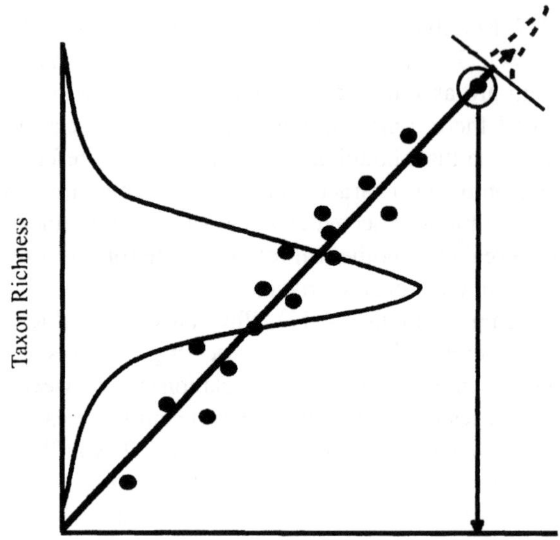

Figure 4-17. Difference in the magnitude and nature of variation among reference sites before and after modeling. Total variation among reference sites biota (solid distribution curve) compared to residual variation (dashed curve). Note circled example site has large deviation from mean in total variation but small residual deviation.

For assessment purposes, only environmental variables that are unrelated to the potential stressor(s) of concern can be used to model the relationships between the biota and the environment of reference sites (Figure 4-18). For example, a model can be based on an attribute of the environment, such as stream flow, that is usually not significantly affected by potential stressors in an assessment of agricultural activity. Then the biota at a test site that is in Reference Condition will be predictable from the stream flow at that site. In this case, environmental variables such as nutrient concentration or riparian vegetation would not be suitable. Similarly, a model relating total phosphorus concentration and the biota at reference sites in lakes must not be subsequently used to predict the biota that should be present at a test site if effects of land use change are being assessed. In that case, one might mistakenly conclude that a test site had the expected reference biota when in fact total phosphorus at the site had increased because of cultural eutrophication to which the biota had responded.

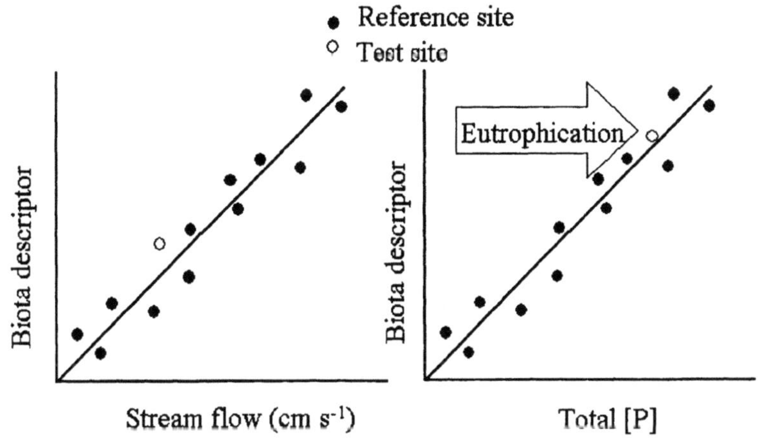

Figure 4-18. The importance of modeling Reference Condition with environmental variables unaffected by stressors. An environmental variable unaffected by organic enrichment, such as flow rate in streams (left), provides a prediction of the test site's biota if it is in Reference Condition. A variable affected by organic enrichment, like total phosphorus in lakes (right), will mistakenly suggest that a lake experiencing cultural eutrophication has the expected reference community.

Thus, assessment of a test site will be more sensitive to deviations from Reference Condition if the expected Reference Condition can be refined using environmental information measured at the site. The more variation among reference sites that can be explained by the environmental descriptors, the more sensitive the assessment of test sites will be (Bailey et al. 1998). This matching of a test site to appropriate reference sites representing the expected Reference Condition based on environmental attributes is done by developing predictive models.

5. HOW TO MODEL VARIATION AMONG REFERENCE SITES

The nature of the data and the manner in which they are treated will partly control the choice of methods for associating environmental features with variation in the Reference Condition. Usually there will be many environmental features that are to be associated with the biotic data, which may be in several forms as described in Chapter 3. The biotic data may be compressed to a univariate response variable such as species richness or a diversity index. All the taxa may be considered at all sites, or sites with similar biota may be grouped as a multivariate response (Figure 4-19).

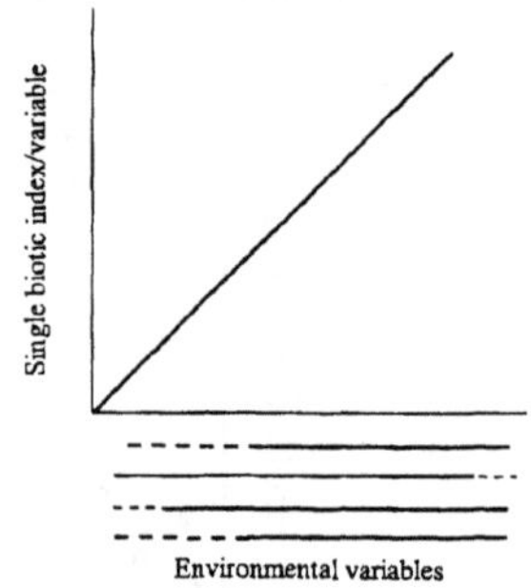

a. Multiple environmental variables x one biotic variable: *Multiple linear regression, general linear model*

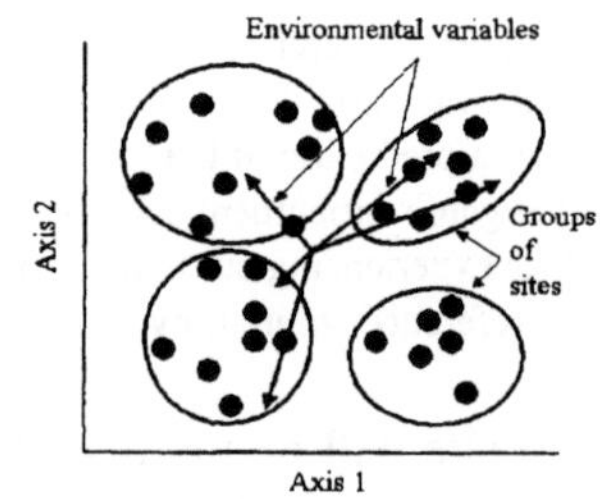

b. Many environmental variables x many groups of sites: *Multiple discriminant function analysis*

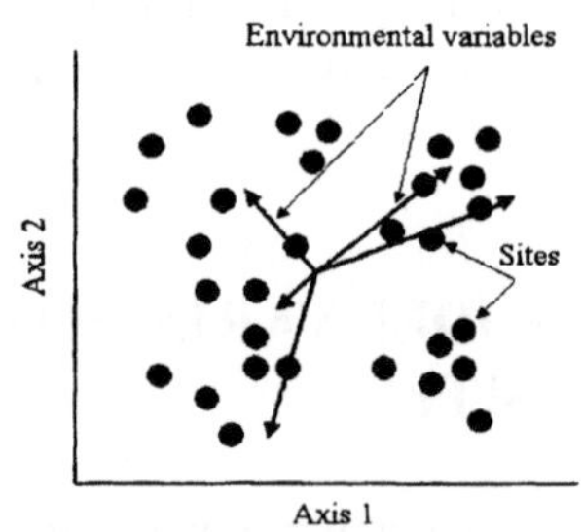

c. Many environmental variables x many taxa or biotic variables at sites: *Correlated ordinations (e.g. CANOCO; PCC)*

Figure 4-19. Analysis options for associating environmental data with corresponding biotic data based on the nature of and format of the data.

5.1 Multiple environmental variables x one biotic variable

With a *single descriptor* of the biota, which can either be univariate (e.g., Taxon Richness, biotic index) or multivariate (e.g., MCDist, an ordination axis score), statistical model-building tools such as multiple regression can be used to establish relationships between the biota and the environment at reference sites (Figure 4-19a). The resulting model can then be used to predict the value of the biota descriptor expected at a test site if it is in Reference Condition, given its environment. The precision of the prediction will be dependent on the length of the environmental gradients in the sample of reference sites and the strength of the association between the gradient and the descriptor. The greater the R^2 of the model, the more the environment explains the variation in the biota (at least as represented by the descriptor). With higher order terms including quadratic and interaction terms, one need not assume linear or even monotonic relationships between the biota and its environment.

5.2 Many environmental variables x many groups of sites

If sites representing the Reference Condition have been drawn together in *groups* with relatively homogeneous biota, using a classification technique, then predictive relationships can be developed between groups of sites and their associated environmental characteristics. Discriminant function analysis (DFA) is an appropriate procedure, as described by Green (1979) and Norris and Georges (1993). The value of a multivariate technique such as DFA compared to univariate approaches is that, when taken alone, each variable may discriminate poorly between the faunally defined groups. When used in combination to define a discriminant function in the direction of maximal variation between groups, the discrimination can be clear (Norris and Georges 1993).

DFA is the approach most frequently used in the development of models for predicting the membership of a site in a previously established classification (e.g., Wright et al. 1984, Moss et al. 1987, Reynoldson et al. 1995, Hawkins et al. 2000, Simpson and Norris 2000). DFA establishes appropriate empirical relationships between the macroinvertebrate groups and environmental variables, so that a subset of the environmental variables adequately predicts the biota. An additional advantage of DFA is that it enables the probability of group membership to be apportioned. As noted earlier, no two sites (or groups of sites) will be identical and the probability of group membership can be used to adjust the prediction of fauna at a test site (Wright 1995, Figure 4-20).

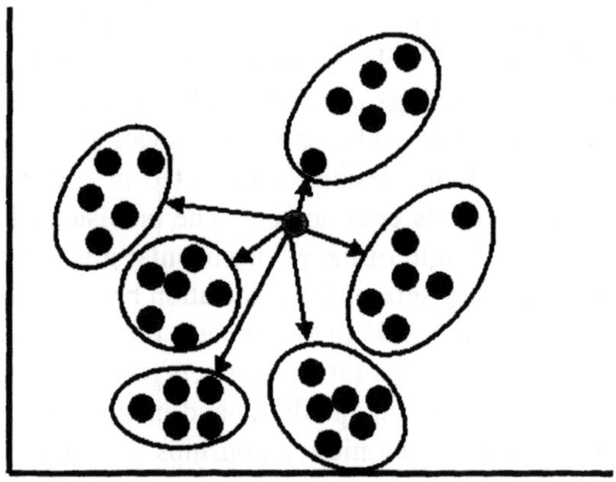

Figure 4-20. Probabilities (shorter arrows are stronger probabilities) of a test site belonging to reference site groups calculated by discriminant function analysis (DFA).

The most common approach is to substitute the environmental data for each site into the discriminant functions to determine the proportion of sites that are misclassified (e.g., Hawkins et al. 2000, Simpson and Norris 2000). Reference sites that were initially classified into faunal groups on the basis of their biota, and then subjected to a DFA to create a model relating group membership to their environment, are re-classified with the DFA model into faunal groups. If there is a strong relationship between the biota and the environment, the sites will be correctly re-classified into their original faunal groups; a weak relationship between biota and environment will lead to a high percentage of re-classification errors.

5.3 Many environmental variables x many taxa or biotic variables at sites

A criticism of the use of DFA is that it requires the formation of groups of sites, based on the biota, that are deemed to represent different states of the Reference Condition. However, in reality the groups are not discrete (see ordination plots above), particularly as the number of reference sites in the dataset increases. More typically they represent a continuum of sites along one or more environmental gradients. Therefore, alternative approaches that acknowledge this may be more appropriate.

Where all sites representing the Reference Condition are considered together, and we want to establish that presence and absence, or abundance of taxa, among sites are related to several measured environmental variables, several methods are available (Figure 4-19c). For all of them, the first step is some type of ordination of the sites using their taxonomic composition, followed by one of several methods to associate the environmental data. So, to begin with, the reference sites are represented by their distribution in multidimensional space using the biota. Then a second ordination, based on the environmental variables is laid over the first ordination. Environmental variables can also be associated with the outputs of ordinations by correlating them with the axes (e.g., Wright et al. 1984).

Ordination is used extensively in studies of benthic invertebrates. Canonical correspondence analysis (CANOCO) is an ordination method commonly used in aquatic ecology to relate the distribution of taxa and their environment (see ter Braak and Verdonschot 1995 for a review). These analyses are appealing because they have been developed from ecological niche theory (Shelford 1911, Odum 1971) where each species thrives best at a particular value (its optimum) and cannot survive when the value is either too low or too high. Each species' occurrence is thus confined to a limited range, its niche (ter Braak and Verdonschot 1995). A species' niche will be described by many environmental variables or dimensions (Hutchinson 1968) and thus multivariate analysis is an appropriate application in this context. CCA is a mathematical formulation for determining the environmental optima, though, like all multivariate methods, it is dependent on the researcher having measured appropriate environmental variables at relevant scales.

Bioenv (Clarke and Ainsworth 1993, Clarke and Warwick 1994) calculates the rank correlations between the dissimilarity matrices derived for ordinations of the biota and the environmental variables; and it selects the best subsets of variables likely to be most important in describing the relationships. Principal axis correlation (PCC procedure, Belbin 1987) is essentially a multiple linear regression that is designed to see how well a set of attributes can be fitted to an ordination space. It determines the linear relationship between any set of ordination scores (axes, based on the biota in this context) and a set of environmental measures. PCC has the advantage that it can be can be applied to axes from almost any sort of ordination (Belbin 1987), but it has the disadvantage that all environmental variables that are correlated in the ordination space of the biota will be highlighted, even if they are correlated with each other.

Any of the procedures introduced above can be used to select environmental variables that help explain variation in the biota at reference sites. In our case studies presented here, we use stepwise multiple linear regression and DFA, but similar models relating the biota to its environment at reference sites could have been obtained with any of these methods

5.4 Selection of predictor variables

Predictor variables, that are likely to account for substantial variability among reference sites and that are unlikely to be affected by human activities, can be suggested from several previous studies (Moss et al. 1987, Reynoldson et al. 1995, 1997, Bailey et al. 1998, Hawkins et al. 2000b, Simpson and Norris 2000). Hawkins et al. (2000b) summarized 11 studies that considered environmental features and aquatic biota and how well these grouped sites, and concluded that a tiered approach that considers small-scale features and then large-scale features provides the best outcomes. Large-scale environmental variables commonly selected include measures of geographic location and proximity in a catchment, such as geographic coordinates, altitude, distance from the source and conductivity or alkalinity as measures of catchment geology. These large-scale variables are common to most models. Subsequently, individual models include local-scale variables such as water depth, stream width and particle size. Selection of these local-scale variables will be dependent on the local area. For example, if sedimentation is likely to be a problem caused by human activities the percentage of silt in the substratum (as chosen by Reynoldson et al 1997) would be inappropriate for testing that area for the effects of land use. Similarly, if land use promoted problems with salinity, conductivity could not be used as a predictor variable.

6. WHY RESIDUAL VARIATION AMONG REFERENCE COMMUNITIES IS IMPORTANT

The Reference Condition Approach accepts, indeed celebrates, that "good is variable" (there is variation among sites in Reference Condition), and in characterizing this variability in the Reference Condition (Section 4.1), we are describing the variability of good communities, or at least those that we have deemed acceptable (see Chapter 3). When we define reference criteria

(Chapter 3), whether they be based on current conditions (pristine, minimally disturbed, or best available) or some historical condition (e.g., streams in pre-European times), there is little to be done about reducing such variability other than to describe and model it. The next phase of the Reference Condition Approach, where we try to explain as much of this variation as possible with environmental variation (Section 4.2), results in collecting together the unexplained variation (the *residual variation*) among reference communities as we prepare to assess a test site.

The residual variation is the uncertainty or imprecision in applying our predictive model to a test site in order to generate an expectation of the biota at the test site if it is in Reference Condition. The less residual variation among reference sites, the more sensitive our assessment of test sites will be, and the smaller the deviation of the test site's biota from the expected Reference Condition we will be able to detect.

Hypothesis tests are not always used in such situations formally, even though the paradigm is useful here. All assessments, whether upstream–downstream, BACI-style designs (before–after control–impact, see Downes et al. 2002), or national, multi-year studies (e.g., Wright 1995), are based on a comparison of a test site, or test sites, to a null hypothesis, or reference site distribution (Chapter 5). Our goal is to decide whether or not the test site could have been part of the reference site distribution. With the Reference Condition Approach, the reference distribution is the distribution of deviations of reference sites from the predicted descriptor(s) of the biota, (i.e., the residual variation). Regardless of the scale of the study or the details of the analysis approach, we want the residual variation to be as small as possible so that we are able to detect, with a good deal of confidence, deviations of test sites from the reference distribution.

In formal statistical terms, as we discuss further in Chapter 5, falsely concluding that a test site is not a member of the reference distribution is a "Type 1 error" or "false positive", while falsely concluding that a site is a member of the reference distribution is a "Type 2 error" or "false negative". For a given, defined "effect size", or deviation from the reference distribution (e.g., three species fewer than expected), the chance of making either kind of error decreases as the residual variation of a model is decreased. Put another way, the more variation among reference communities in Taxon Richness, for example, that is explained by environmental descriptors, the less the chance that one will make a mistake in concluding that a test site either is or is not a member of the reference distribution (Figure 4-21).

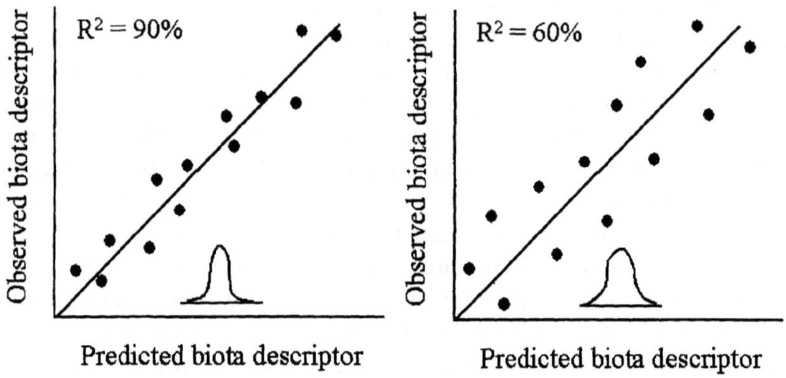

Figure 4-21. Residual variation among reference communities. A smaller amount of residual variation in the data on the left will allow a more sensitive test of deviation of a test site from Reference Condition.

6.1 Linear model residuals

No matter how well variability in the Reference Condition is modeled, some residual variation will remain unexplained. A linear model approach can be used to explain variation among reference communities in a single descriptor of the biota. It illustrates some of the potential trouble spots in creating and applying the predictive model. Regardless of the number or types of predictors used in the model, it is always possible to plot the observed value of the response variable on the y-axis, and the expected or predicted values of the response variable on the x-axis, along with a line of equality indicating perfect prediction of the response variable in the model. As illustrated in Figure 4-21, this provides an easy graphical way to compare predictive models created for different response variables, or different predictive models created for the same response variable. The R^2 value is 100% minus the ratio of the residual variance to the total response variable variance.

Trouble spots in the predictive model can be detected with these plots. In some cases, predictive models do a better job in some parts of the response variable range than others (Figure 4-22, left), or there is a systematic bias in the prediction depending on the predicted value (Figure 4-22, right).

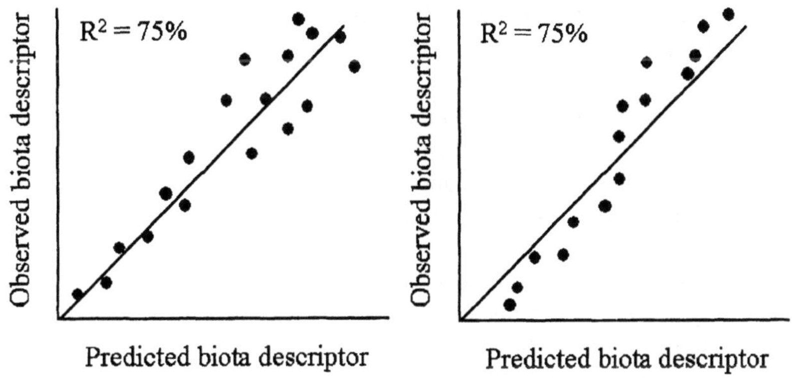

Figure 4-22. Detecting problems with predictive models by analyzing residual variation. Residual variation may be either randomly (left) or systematically (right) distributed in heterogeneous fashion.

Distribution curves of residual variation can also show non-normal asymmetry around the mean or median, which will have ramifications for the sensitivity of the test (Figure 4-23). If the variation is asymmetric, then the sensitivity will be greater in the direction of less variation.

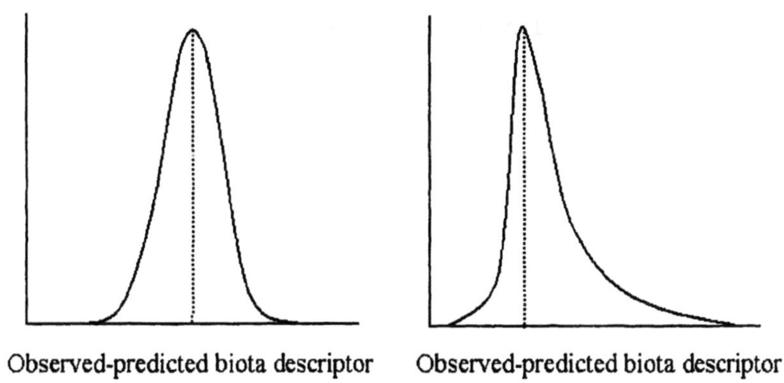

Figure 4-23. Distribution of residual variation. Residual variation may be approximately normal or at least symmetric (left), or non normal (right). In the right-hand distribution, it will be easier to detect a small deviation from Reference Condition in the negative than the positive direction.

6.2 Ordination score residuals: the BEAST approach

In some applications of the Reference Condition Approach, groups of sites with similar fauna are formed with a classification technique. DFA, with the faunal groups as DFA groups and environmental descriptors as response variables, is used to explain community variation and build a predictive model (Section 4.2). The way reference sites vary in distance from the center of their faunal group is quantified in multivariate space by either a distance measure (e.g., Bray–Curtis distance, which we call "BEAST MCDist") or ordination score (Figure 4-24).

There are two places where such a model incorporates variability among reference sites. First, the environmental variables clearly do not perfectly explain the variation among the groups. In other words, the environmental variables will not perfectly predict the correct group membership of a given site. Second, there is some variation among reference communities within each of the faunal groups (Groups A and B in Figure 4-24). So even if the environment predicts perfectly the faunal group into which a site falls, there is some variation in biota among the sites within the group. Therefore there is also some variability in the nature of the predicted biota of a site whose environment suggests that it is a member of a particular faunal group. In BEAST (Reynoldson et al. 1995) and other similar approaches, it is this variation that is captured in the residual community variation.

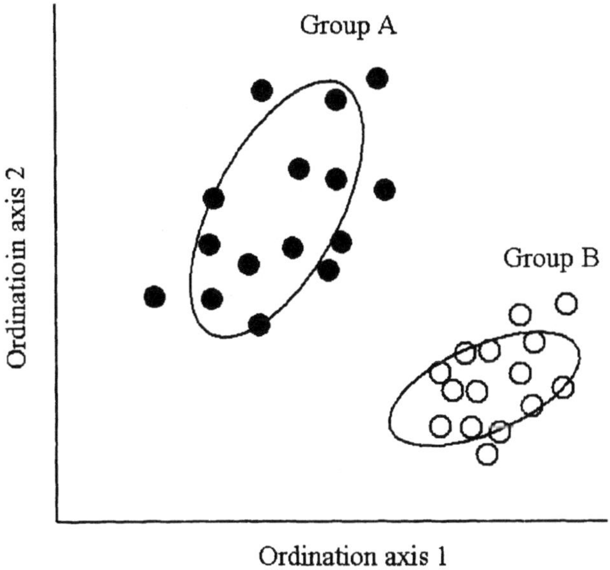

Figure 4-24. Residual variation in multivariate space. The variability within each faunal group is not explained by the predictive model.

This technique does a good job of showing which of the faunal groups are better or worse for detecting a test site's deviation from Reference Condition (that is, which groups have more or less residual variation around their group centroids). If a site is predicted to be a member of Group B in Figure 4-24, we can make a much more precise judgment about its membership than if it is predicted to be in Group A, because there is less variability among communities in Group B than Group A. But the technique does not identify the groups that are more or less predictable from environmental characteristics of a site. No part of Figure 4-24 enables us to calibrate our decision or expectations of error with the probability of a site's membership in the faunal groups, although this is evident from the probability of group membership from the DFA model.

6.3 Observed and expected numbers of taxa

Like BEAST, RIVPACS (Wright 1995, Hawkins et al. 2000b) and AUSRIVAS (Simpson and Norris 2000) also use faunal groups and DFA to create a predictive model of communities based on environments. But unlike BEAST, after the DFA modeling, the expected number of taxa at a site is calculated based on the site's probability of membership in each group and the percent occurrence of each taxon within the group. The deviations between the observed and expected number of taxa can be illustrated using the graphical techniques already described (e.g., Figures 4-21).

Commonly in AUSRIVAS and RIVPACS applications, the ratio of observed number to expected number of taxa (O:E) for each reference site is plotted in a frequency distribution. If there is a perfect relationship between biota and environment, and the groups are perfectly homogeneous in their composition (i.e., every taxon present in a group is present in all sites within the group), then all of the O:E values will be unity. The degree of variability in the distribution of O:E values around unity will indicate both an imperfect relationship between biota and environment, and heterogeneous community composition within faunal groups. As with other analyses of residual variation, plots of observed *versus* expected numbers of taxa with little deviation about the line of equality (e.g., Figure 4-27) or equivalently, O:E frequency distributions that are narrower (e.g., Figure 4-26) are better able to detect deviations of a site from Reference Condition. Asymmetric distributions of the O:E distribution indicate greater ability to detect deviations from reference in one direction than the other.

An extension of the DFA model has been used in some applications of the Reference Condition Approach (e.g., RIVPACS, AUSRIVAS, Wright 1995, Simpson and Norris 2000), to produce a site-specific probability of each taxon in the model occurring at sites of concern. The procedure, outlined in Table 4-3, is as follows. As previously described, the DFA model uses the environment of a test site to produce a set of probabilities of its membership in each of the faunal groups. Each probability of group membership is multiplied by the proportion of sites within the group at which a particular taxon is found (i.e., its frequency of occurrence). For each taxon, the sum of these products across all groups is the expected probability of finding the taxon in the test site if it is in Reference Condition (68% in the example in Table 4-3). The sum of all of these probabilities across all taxa is the expected number of taxa at the test site if it is in Reference Condition.

Table 4-3. Calculation of probability of occurrence of a taxon at a test site based on a DFA model relating the test site to reference sites grouped using their biota and using the environmental features of the site (after Wright 1995).

Classification group	Probability that site X belongs to group	Frequency of taxon in group (%)	Contribution to probability that taxon will occur at site X
A	0.7	80	56
B	0.2	50	10
C	0.1	20	2
Probability that taxon will occur at test site X if test site is in Reference Condition			**68%**

The number of observed taxa (O) can be compared with the number expected as calculated from the model (E), and the outcome (O:E) is a measure of biological condition of the site. The ratio should be unity for a perfect model assessing a site that has no damage. In practice, models are not perfect, and model error is usually estimated by comparing the number of taxa observed at reference sites to the number expected from the model, and generating a distribution of reference site O:E ratios. This population of O:E ratios describes the distribution of errors in predicting Taxon Richness and is used to determine if the O:E ratio of a test site is different from that expected from model error alone (Hawkins et al. 2000a).

When making comparisons for assessment purposes, the absence of a taxon that had a very low probability of occurring at a test site conveys almost no information about the condition of the site, since it had a high probability of absence by chance alone. Thus, in the Australian application of RIVPACS, only those taxa with >50% chance of occurring are used in calculating the O:E ratios (Simpson and Norris 2000). This issue has been rigorously tested by Hawkins et al. (2000a) who conclude that models using only taxa with >50% chance of occurring are more robust than those using all taxa.

This extension of the DFA approach is valuable in several respects. It provides a taxon by taxon prediction of what we expect to see at a site if it is in Reference Condition, given its environment. It also explicitly incorporates uncertainty into the prediction of biota at a test site based on its environment, including the probability of group membership and probability of occurrence of taxa in sites which are members of the group.

7. CASE STUDIES: MODELING AND EXPLAINING VARIATION AMONG REFERENCE SITES

For the case studies, we constructed predictive models using some of the approaches described above including multiple regression (with single descriptors of the biota) and DFA (with groups of sites defined by their fauna) and assess how well they explain variation in the Reference Condition and how much variation is left unexplained. The following models are created and evaluated.

1. *Multiple regression* models using many environmental variables measured in each of the case studies to predict MCDist (Bray–Curtis distance from the median reference community):
   (a) Evaluation with R^2 of the regression model and residual variation,
   (b) Observed versus expected plots for MCDist.

2. *DFA* on groups of sites based on similarity of their fauna. We used stepwise DFA to select a subset of predictor variables related to groups of reference sites in the following case studies.
   (a) Evaluation by misclassification error of reference sites using predictor variables to their pre-determined fauna based groups, but recognizing that the number of groups will be related to the error.
   (b) BEAST models, evaluated by the variation of the biota around the group centroid.
   (c) Observed versus expected plots of the number of taxa observed belonging to the prediction list of the model against the number expected to occur with a probability >50%, evaluated by a comparison of the ranges of O and E, the slope and the R^2 of the relationship and the intercept.
   (d) AUSRIVAS/RIVPACS models using the distribution of O:E ratios in a histogram to evaluate the mean, spread and possible skew.

7.1 Near shore Environment of North American Great Lakes

A multiple regression model was fitted to the reference sites in the Great Lakes using the multivariate distance of the community from the median reference community (MCDist) as a descriptor of the biota. The environmental variables explained a modest amount of the variation in MCDist (R^2 = 26%). Calcium oxide and magnesium oxide in the sediment were important predictors in the MCDist model. (We do not consider that this is a causal relationship; rather it is a convenient, empirical way of making predictions.) By plotting the observed *versus* expected MCDist values from the model, we can see that it does quite a poor job of predicting communities that are close to the median community (MCDist < 0.5), but does moderately well at identifying communities that, based on their environment, are substantially different from average (MCDist > 0.8; Figure 4-25).

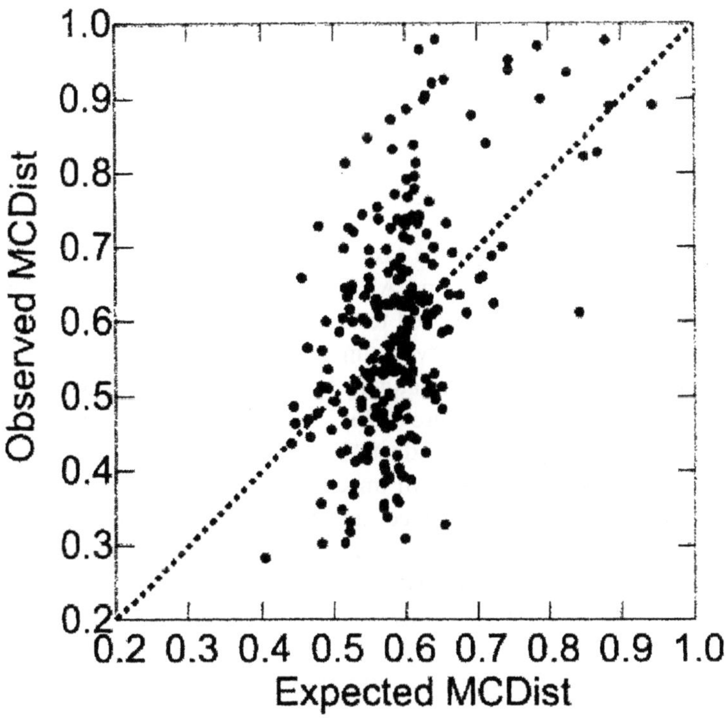

Figure 4-25. Residual variation in MCDist among Near shore Great Lakes reference communities following application of the predictive model.

The DFA of Great Lakes reference sites was moderately successful. About half (52%) of the reference sites used to build the model were correctly re-classified (using the cross-validation procedure) into their original faunal group using a suite of environmental variables that included a major influence of location (latitude, longitude) and depth (Table 4-4).

Table 4-4. Re-classification of Near shore Great Lakes reference sites using the discriminant functions from environmental variables. 48% of reference sites were misclassified with the DFA model.

	Number of Observations Classified into Group										
Group	**1**	**2**	**3**	**4**	**5**	**6**	**7**	**8**	**9**	**10**	**Total**
1	**16**	5	4	0	0	0	1	1	1	0	28
2	0	**13**	0	1	1	2	0	1	2	0	20
3	2	1	**10**	0	0	0	0	0	0	0	13
4	6	1	3	**7**	7	0	0	0	0	0	24
5	0	6	1	8	**20**	4	0	1	0	0	40
6	1	2	3	1	1	**4**	0	2	1	0	15
7	0	2	0	0	0	1	**15**	2	7	6	33
8	0	2	0	0	5	0	0	**9**	2	2	20
9	1	1	0	0	0	0	2	0	**9**	3	16
10	0	0	0	0	0	1	4	2	7	**27**	41
Total	26	33	21	17	34	12	22	18	29	38	**250**

When the probabilities of the taxa occurring at each site (using a DFA based model as in RIVPACS and AUSRIVAS) were summed, an expected number of taxa could be compared to the number of taxa actually observed. Calculation of the ratio of observed number of taxa and expected number of taxa (O:E values) showed a relatively broad distribution skewed towards those smaller than unity (Figure 4-26). The sites with O:E values <0.75 would usually be reviewed and if determined to have problems with sampling or a previously unknown impact may be omitted from the model. Plotting observed *versus* expected number of taxa from this modeling showed that sites where a small number of taxa were expected tended to have even fewer than predicted, whereas those with more predicted taxa tended to be richer than expected (Figure 4-27).

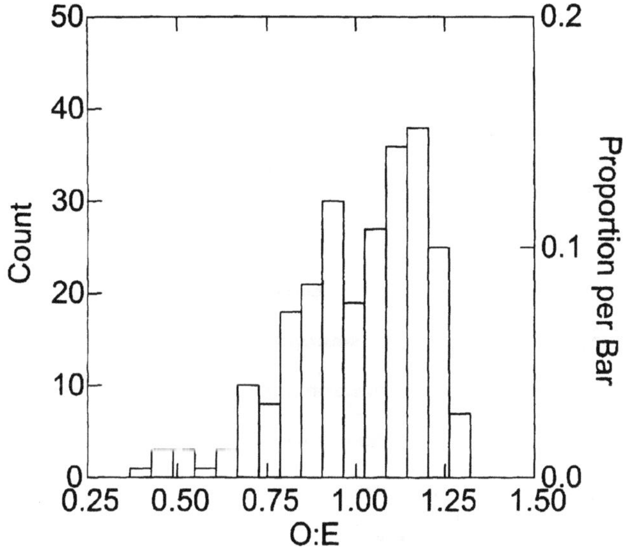

Figure 4-26. O:E ratios (ratio of observed to expected number of taxa) for reference sites from the Near shore Great Lakes based on the DFA extended predictive modeling.

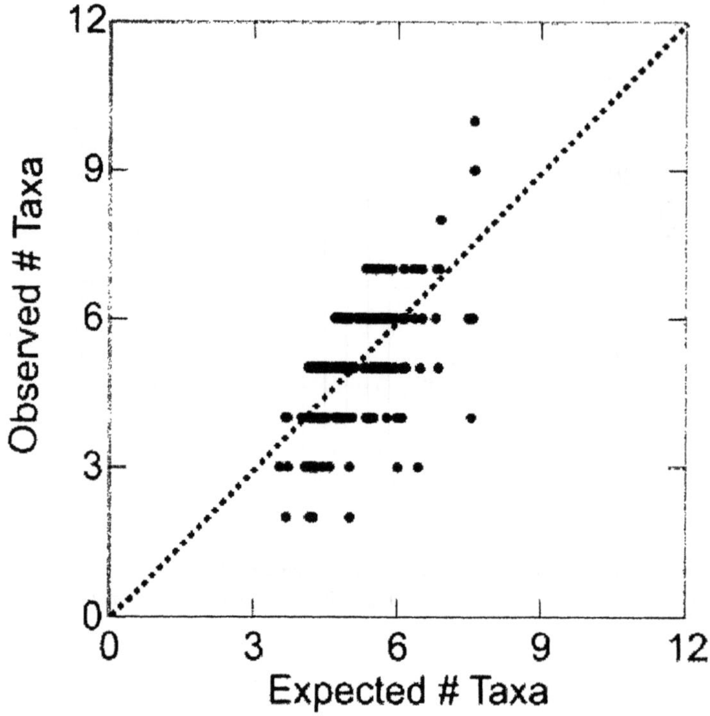

Figure 4-27. Observed *versus* expected number of taxa for reference sites from the Near shore Great Lakes based on the DFA extended predictive modeling. $R^2 = 0.37$.

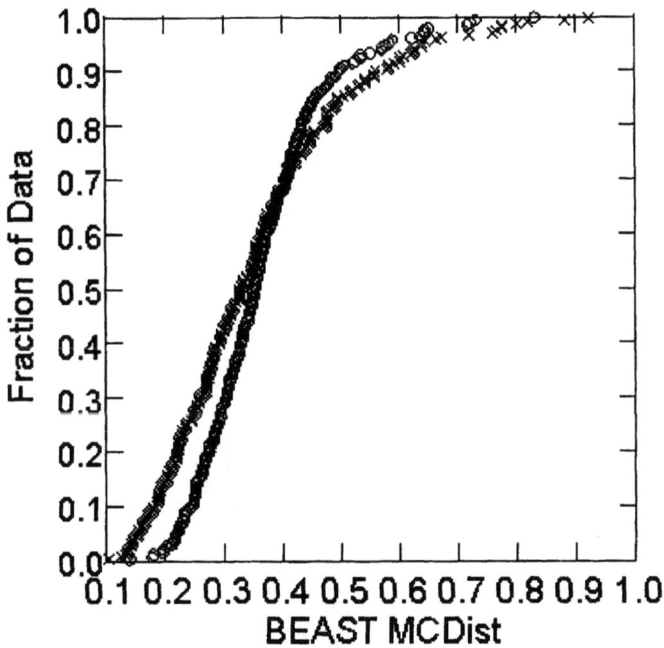

Figure 4-28. Distribution of Bray–Curtis distances between reference communities and the median community of their cluster group for Great Lakes (X) and Fraser River Basin (O) studies.

Residual variation in the BEAST approach to assessment really amounts to the variability of community composition within each of the cluster groups used to build the DFA predictive model, as portrayed graphically with ordination (Figure 4-4). This variability can be quantified as the Bray–Curtis distance between each reference community and the "center" (or centroid) of its group, a value we call "BEAST MCDist", since it is like calculating MCDist for each of the faunal groups. We calculated BEAST MCDist values for each reference site and determined their frequency distribution (Figure 4-28). For the Great Lakes study, 90% of the reference sites have a BEAST MCDist value of about 0.6 or less. (A BEAST MCDist value of 0 would mean that the community is identical to the median or typical community of its group.) When we overlay the distribution of the BEAST MCDist values from the Fraser River study (Figure 4-28) we see that the Fraser River faunal groups were more tightly defined (i.e., had smaller maximum values of BEAST MCDist) than the Great Lakes groups, although the median BEAST MCDist value was about 0.35 for both studies.

7.2 Fraser River Biomonitoring Program

As with the Great Lakes data, a multiple regression model was fitted to the reference site data with MCDist as the univariate response variable. The environment explained a moderate amount of variation in MCDist (R^2 = 22%). Altitude and riparian vegetation cover were important aspects of the environment in the MCDist model. As with the Great Lakes model, biota very different from the median community were better predicted than those with small MCDist values (Figure 4-29).

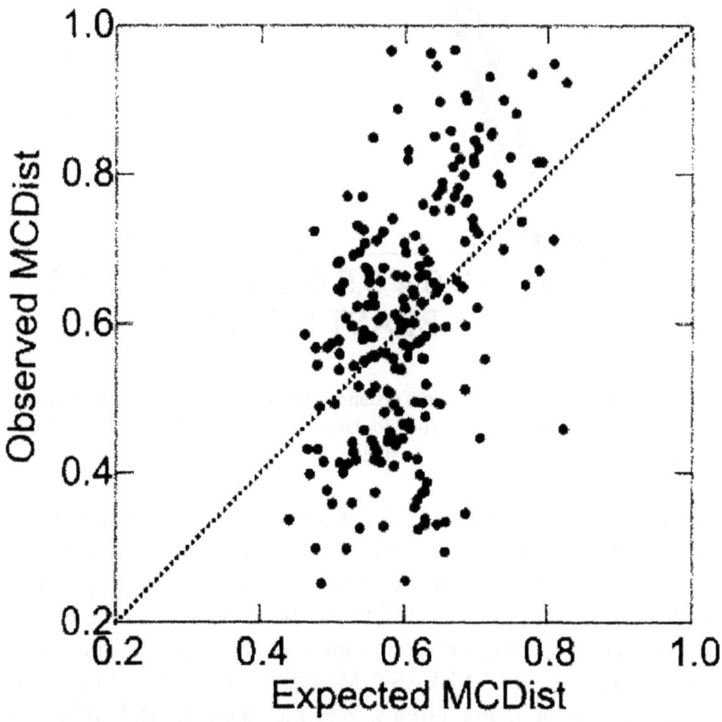

Figure 4-29. Residual variation in MCDist among Fraser River Biomonitoring Program reference communities following application of the predictive model.

DFA of reference sites from the Fraser River Biomonitoring Program had similar success to the Great Lakes model, although fewer groups were used, thus providing fewer opportunities for errors. Reference sites used to build the DFA model were reclassified with a 54% success rate (Table 4-5). Important environmental variables in the DFA model included bank width of the stream site and alkalinity, as well as the degree to which fine sediment (<2 mm) was found in the substrata surrounding larger particles (>2 mm).

Table 4-5. Re-classification of Fraser River reference sites using the discriminant functions from environmental variables. 46% of reference sites were misclassified with the DFA model.

	Number of Observations and Percent Classified into Group						
From Group	**1**	**2**	**3**	**4**	**5**	**6**	**Total**
1	**22**	7	8	4	2	2	**45**
2	4	**13**	0	1	0	0	**18**
3	13	1	**19**	6	5	9	**53**
4	7	7	6	**30**	2	3	**55**
5	2	5	1	4	**8**	1	**21**
6	2	1	3	0	1	**18**	**25**
Total	**50**	**34**	**37**	**45**	**18**	**33**	**217**

Predictions of the number of taxa at each site (as in RIVPACS and AUSRIVAS) showed a more uniform distribution of O:E values about unity (Figure 4-30) than in the Great Lakes study (Figure 4-23). Again, the single site with a very low O:E value would automatically be reviewed. However, the review might also be extended to sites with O:E values <0.75 and >1.25. This is not to suggest that these sites would automatically be excluded. Their omission would only be justified if it were determined that either sampling, or some external impact had contributed to variability that was not part of the Reference Condition.

The observed *versus* expected number of taxa from the RIVPACS-type DFA model also showed less pattern than the Great Lakes in the deviation of observed and expected number of taxa with varying predicted values (Figure 4-31). The model was slightly better than the Great Lakes model.

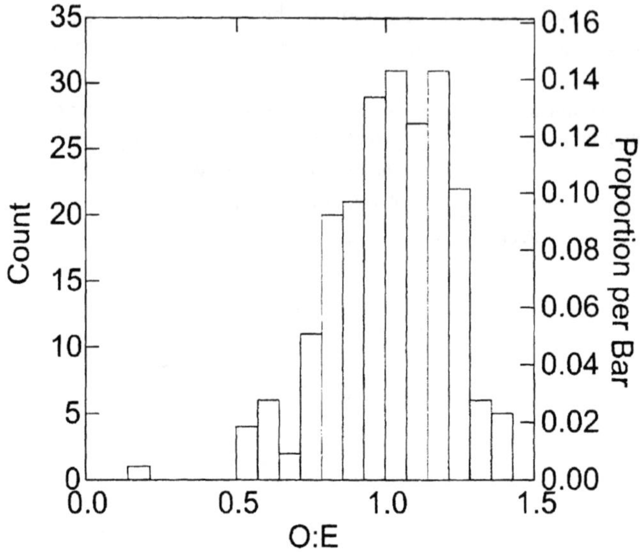

Figure 4-30. O:E ratios (ratio of observed to expected number of taxa) for reference sites from the Fraser River Bioassessment Program based on the DFA extended predictive modeling.

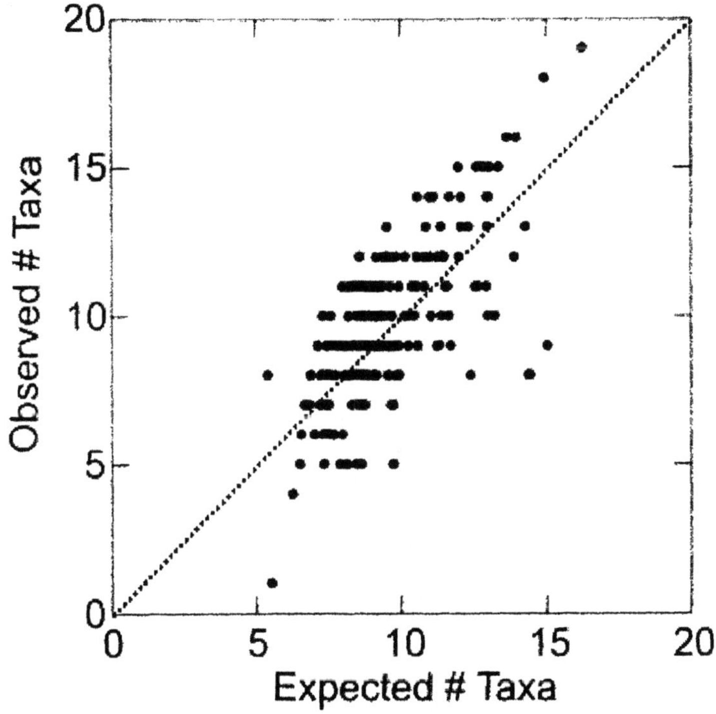

Figure 4-31. Observed *versus* expected number of taxa for reference sites from the Fraser River Bioassessment Program based on the DFA extended predictive modeling. $R^2 = 0.44$.

As previously discussed, the BEAST MCDist residuals for the Fraser River study showed a less variable, tighter distribution of sites within their faunal groups as compared to the Great Lakes study (Figure 4-28). This would indicate greater sensitivity in the Fraser River study relative to the Great Lakes in detecting deviations from Reference Condition when using the BEAST assessment approach.

7.3 Spring and Autumn Assessment of Streams in the Australian Capital Territory

Two separate multiple regression models were fitted to the Spring and Autumn reference site data from the Australian Capital Territory with MCDist as the response variable. In the Autumn sample, a very small amount of variation in MCDist was explained by variation among environments of reference sites (R^2 = 18%; primary predictors included altitude and % gravel in the edge). For the Spring sample, a moderate amount of variation in

MCDist was explained by the model (R^2 = 33%; trailing edge vegetation and proportion of riffle habitat).

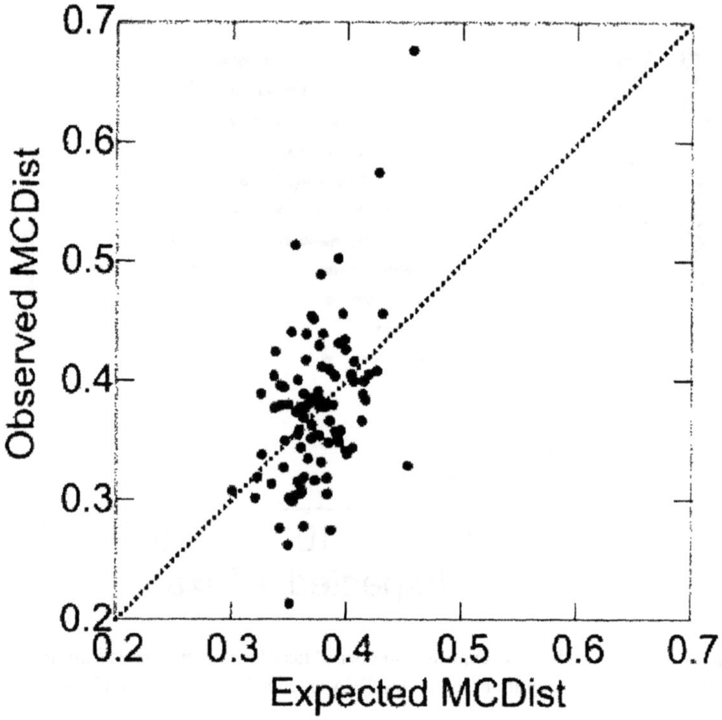

Figure 4-32. Residual variation in Autumn MCDist among Australian Capital Territory reference communities following application of the predictive model.

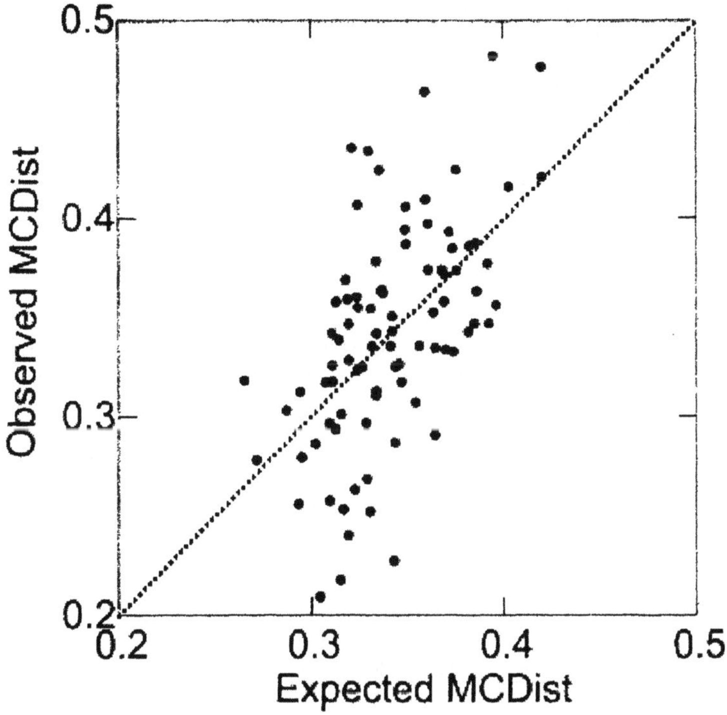

Figure 4-33. Residual variation in Spring MCDist among Australian Capital Territory reference communities following application of the predictive model.

There was greater success in explaining variation among faunal groups with the DFA of Australian Capital Territory reference sites than with either the Great Lakes or Fraser River studies. Both the Autumn (Table 4-6) and Spring (Table 4-7) samples had only 26% misclassification rates when reference sites used to build the DFA model were reclassified using only the environmental data. Important environmental variables in the Autumn DFA model included latitude, longitude, bank width, and pH; sites in the Spring model were classified primarily on the basis of longitude, altitude, and conductivity. However, the smaller number of groups in these models (4 and 5 compared to 10, Great Lakes and 6, Fraser River) would account for some of this difference.

Table 4-6. Re-classification of Australian Capital Territory Autumn reference sites using the discriminant functions from environmental variables. 26% of reference sites were misclassified with the DFA model.

	Number of Observations and Percent Classified into Group				
From Group	**1**	**2**	**3**	**4**	**Total**
1	12	6	2	3	23
2	1	3	0	1	5
3	3	5	26	3	37
4	2	8	2	16	28
Total	**18**	**22**	**30**	**23**	**93**

Table 4-7. Re-classification of Australian Capital Territory Spring reference sites using the discriminant functions from environmental variables. 26% of reference sites were misclassified with the DFA model.

	Number of Observations and Percent Classified into Group					
From Group	**1**	**2**	**3**	**4**	**5**	**Total**
1	9	0	0	6	1	16
2	2	11	3	1	1	18
3	1	0	5	0	0	6
4	4	0	0	21	0	25
5	0	1	1	2	19	23
Total	**16**	**12**	**9**	**30**	**21**	**88**

The number of taxa predicted at each site (as in RIVPACS and AUSRIVAS) showed a similar, relatively uniform distribution of O:E values about unity in both the Autumn and Spring samples from the Australian Capital Territory (Figures 4-34, 4-35). As in the previous models the seven Spring and nine Autumn sites with O:E values <0.8 or >1.2 would usually be reviewed. Based on these distributions predictions from the models would be expected to be less biased and more sensitive with the Spring model than the Autumn model because the O:E frequency distribution is more evenly distributed around 1.0 and slightly narrower than the Autumn model. Both are likely to be better than the Great Lakes and Fraser River models.

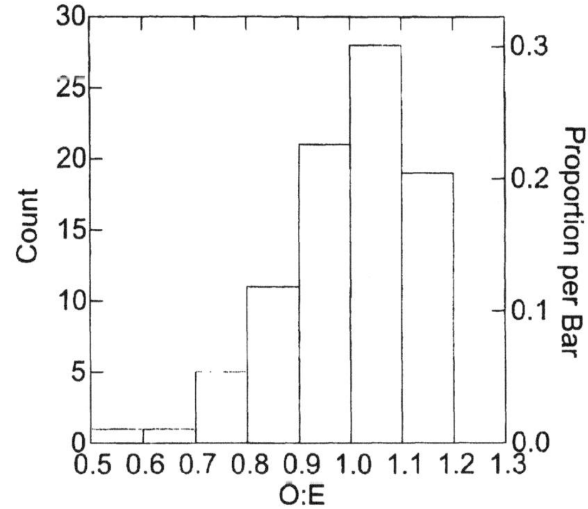

Figure 4-34. O:E ratios (ratio of observed to expected number of taxa) for reference sites from the Australian Capital Territory (Autumn sample) based on the DFA extended predictive modeling.

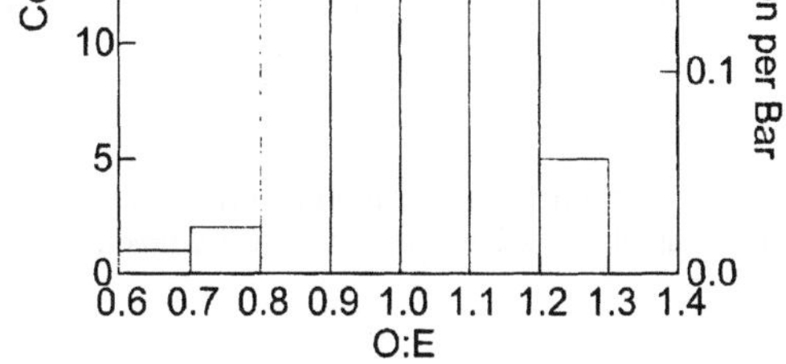

Figure 4-35. O:E ratios (ratio of observed to expected number of taxa) for reference sites from the Australian Capital Territory (Spring sample) based on the DFA extended predictive modeling.

Although a given community tended to have more taxa in the Autumn than the Spring samples, the observed *versus* expected number of taxa from the extension of the DFA model showed similar magnitude of deviation between the observed and expected, regardless of the number of taxa predicted (Figure 4-36, 4-37). Based on their higher R^2, and the pattern of observed and expected taxa, both of these Australian Capital Territory models would yield predictions closer to observed taxonomic composition than the Great Lakes or Fraser River models and the Spring model performs better than the Autumn model.

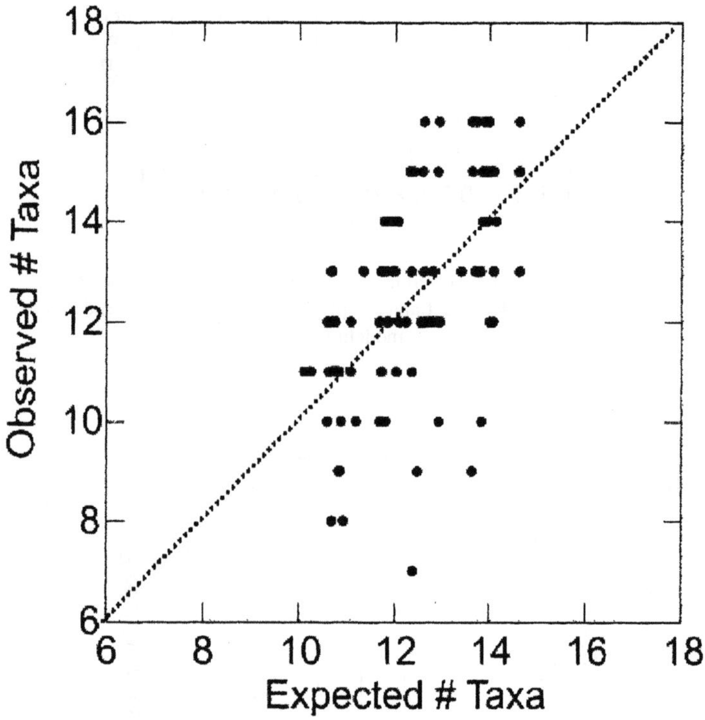

Figure 4-36. Observed *versus* expected number of taxa for reference sites from the Australian Capital Territory (Autumn sample) based on the DFA extended predictive modeling. $R^2 = 0.30$

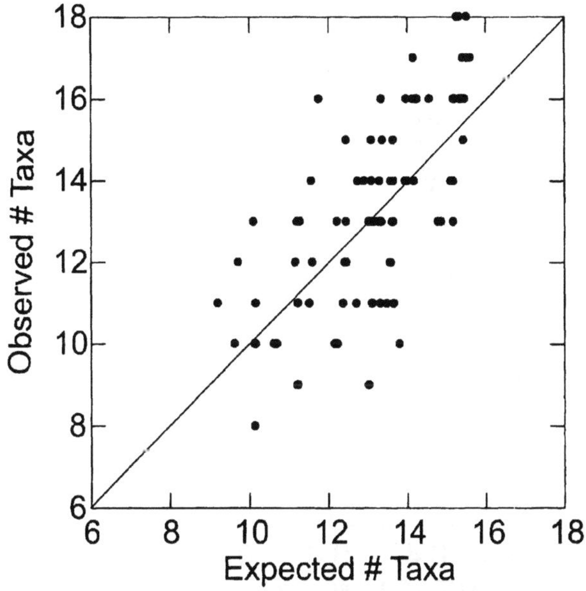

Figure 4-37. Observed *versus* expected number of taxa for reference sites from the Australian Capital Territory (Spring sample) based on the DFA extended predictive modeling. $R^2 = 0.49$.

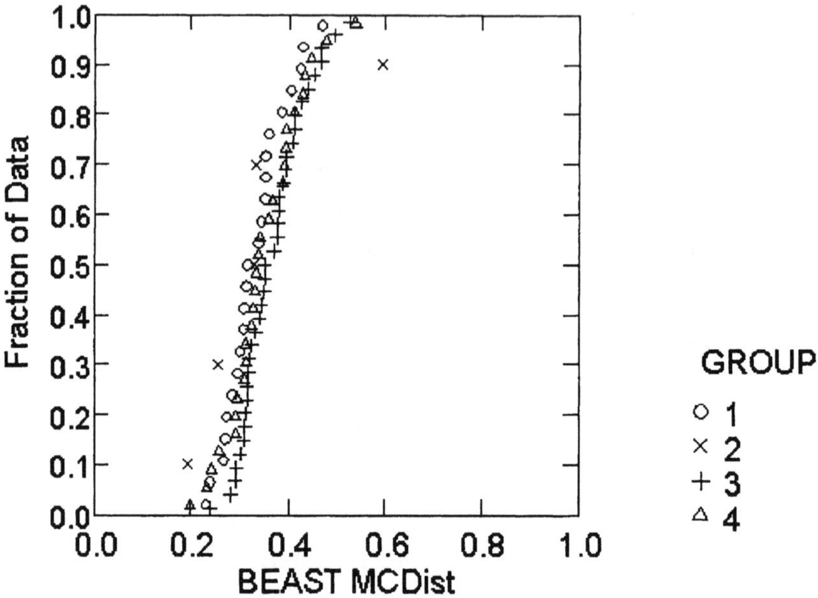

Figure 4-38. Distribution of Bray–Curtis distances between Australian Capital Territory Autumn reference communities and the median community of their cluster group (see Figure 4-15).

All sites within a faunal group from the Autumn sample period had similar variability in composition (Figure 4-38), and both Autumn and Spring (Figure 4-39) communities were much more similar to the median of their faunal groups than sites in either the Great Lakes or Fraser River study (Figure 4-28). It is clear in looking at the distribution of BEAST MCDist values from the Spring sampling period that faunal Group 3 consists of a more homogeneous set of communities (i.e., they tend to be closer to their median or average community of the group).

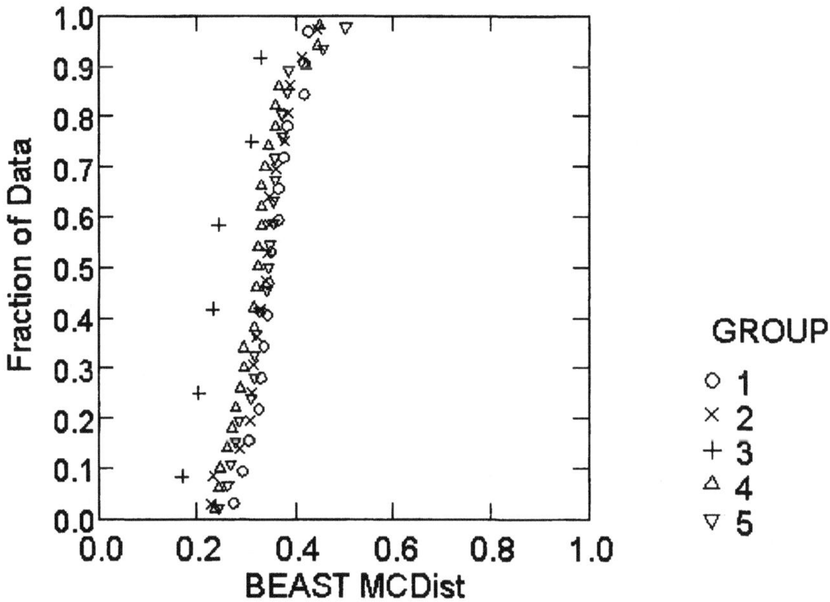

Figure 4-39. Distribution of Bray–Curtis distances between Australian Capital Territory Spring reference communities and the median community of their cluster group (see Figure 4-16).

With each of the case-study datasets, we have demonstrated three approaches (multiple regression with a single biota descriptor, DFA with calculated variability within faunal groups, and DFA with calculation of the identity and number of expected taxa) to building a model that relates the biota at reference sites to their environment. In some cases (e.g., modeling MCDist in a multiple regression with the Australian Capital Territory Autumn sample), the modeling only explained a very small amount of variation in reference biota, and thus will gain us little sensitivity when we actually assess test sites (Chapter 5). In other cases (e.g., DFA followed by calculation of O:E values for the number of taxa in the Australian Capital Territory Spring sample) about half of the natural variation among reference sites was explained and a significant enhancement in the sensitivity of the assessment can be made. Our point here is not to suggest which of these particular strategies will always work the best; the most effective will vary among different studies in different contexts. Rather, we have demonstrated the utility of the modeling procedure itself in sensitizing the eventual assessment decision-making as described in the next chapter.

8. POINTS TO REMEMBER

— Biota and environmental features vary naturally among reference sites. This variability, which is fundamental to the Reference Condition Approach, is described in the first step of the approach.

— Several decisions will be needed on the treatment of the data (e.g., elimination of rare taxa, presence or absence versus abundance, transformation) before proceeding with analyses.

— Simple histograms, box plots, bivariate scatter plots, and multivariate statistical techniques of classification and ordination are commonly used

— methods for describing variation among reference sites.

— Constructing a predictive model that accounts for some of the biota variation among sites in Reference Condition will improve assessment of test sites.

— Several methods are available for modeling the Reference Condition. The choice of method may depend on the biotic descriptor and whether the reference sites are classified into groups with relatively homogeneous biota.

— Predictor variables used for modeling variation in the Reference Condition must be unrelated to the stressors likely to occur in the study region. Several studies have found that large-scale variables such as geographic location and catchment area are useful predictors.

— Reference sites that fall near the boundaries of the model distribution should be reviewed but only removed if it can be clearly demonstrated that the data have suffered from poor sampling (over or under), or some previously unknown impact.

— Even the best models will have some unexplained, or residual, variation. It should be considered to see if it affects the model quality and the accuracy of predictions possible from the models.

Chapter 5

DECISION-MAKING

Is the site in Reference Condition?

1. WHY DO WE NEED DECISION RULES ON PASSING AND FAILING TEST SITES

In science, a standard approach to hypothesis testing is to establish null and alternative hypotheses, collect data, and then determine the probability that those data would be collected if the null hypothesis were true (Quinn and Keough 2002). If the probability is small enough, we reject the null hypothesis and conclude that the alternative hypothesis is supported.

With the Reference Condition Approach, our null hypothesis is that a test site (say, a lake site at a cooling water outfall from a generating plant) or a set of test sites (say, streams in clear-cut forest) is in Reference Condition. Our uncertainty about the condition of the site(s) is proportional to our uncertainty about the distribution of the Reference Condition, and the biota and their environment at the test site (Figure 5-1).

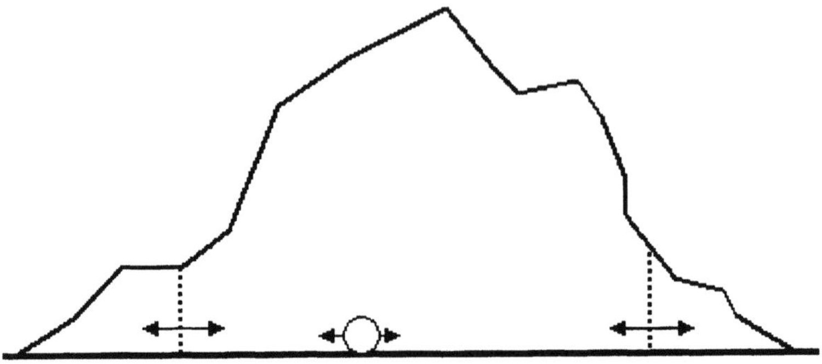

Figure 5-1. Uncertainty in deciding whether or not a test site is in Reference Condition. The distribution of some descriptor of the biota or environment of reference sites is shown, as well as the value of the descriptor for one test site.

Uncertainty in hypothesis testing is quantified by error probabilities. We *sample* both the reference and the test sites, and therefore *estimate* (rather than know for certain) the values of the biota and environment descriptors that we measure. It is possible that if we establish a decision point and determine on which side of the decision point a test site is, we may mistakenly judge the test site to be in Reference Condition (a Type II error) or not in Reference Condition (a Type I error). We will never know whether we have made either a Type I or Type II error, or whether we have made an error at all, but we *can* quantify the probability of the two types of error.

Obviously, we would like to minimize the chances of making either error, but to some degree they trade off with one another. Increasing the chance of mistakenly failing a site in Reference Condition will reduce the chance of mistakenly passing a site that is not in Reference Condition, and *vice versa*. Describing variation in the Reference Condition by modeling it (Chapter 4) will reduce the likelihood of making both types of errors. But we must still determine their relative importance in any assessment study.

The relative importance of decision errors will be based, in part, on the responses and costs associated with the assessment. If environmental protection is a primary concern, we should err on the side of failing sites that may not be damaged. This approach can be justified because the consequences of damage can be serious and costly. Also, management protocols are likely to include review of the data and more detailed site assessment before intervention, and those steps are likely to be inexpensive compared to the possible subsequent need for rehabilitation. The converse may be true if costly mitigation is envisaged. In such cases we may want to err on the side of passing sites that could be in less than Reference Condition, if there is a high public cost involved in taking action or balancing where mitigation resources are to be used.

2. HOW ARE DECISION RULES ARRIVED AT?

In the past, decision rules for biological assessments were usually based on what amounts to a consideration of Type I error. The distribution among reference sites of a certain descriptor of the biota (including residual variation if modeling has been done as in Chapter 4) is considered to be the null hypothesis distribution, and test sites that are in one or both of the tails of the distribution (depending on the nature of the descriptor) "fail" (Figure 5-2a). This is very much akin to the standard approach to statistical hypothesis testing described at the start of this chapter. We determine the chance of a test site having a particular value for the descriptor if it is in Reference Condition. If that probability is below some threshold (i.e., beyond some decision point on the null hypothesis distribution), then we conclude that the test site is not in Reference Condition.

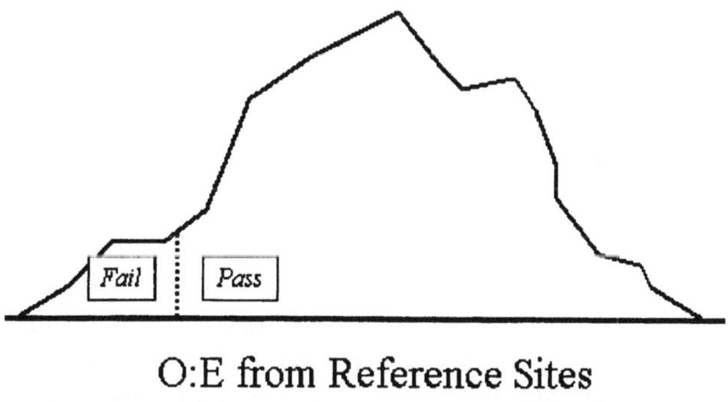

Figure 5-2. A pass–fail decision point (dotted line) based on variability in the Reference Condition (the null hypothesis distribution) and the desired Type I error probability. Test sites with an O:E value less than the 10th percentile from the reference distribution will fail, but with this decision point, 10% of the reference sites would mistakenly be failed.

The US EPA and some others (e.g., Bailey et al. 1998, Linke et al. 1999) have taken a statistically liberal (but environmentally conservative) approach when defining such decision points (Barbour et al. 1999, Gerritsen et al. 2000). They use the 25th percentile of reference site variability as the decision point for passing or failing test sites (that is, deciding the test site is or is not in Reference Condition). Put another way, they define a Type I error probability as 25%; that is, one quarter of the sites that are in Reference Condition would be mistakenly identified as deviating from Reference Condition.

A similar approach has been used to determine decision points for pass and fail for predictive models in Britain (Wright 1995, Clarke 2000) and Australia (Simpson and Norris 2000), although a smaller Type I error probability is used for those tests. In each case, observed to expected ratios (O:E values) are calculated for various indices of biological quality, such as number of taxa, average score per taxon (ASPT), Biological Monitoring Working Party (BMWP), using predictive models for different regions, seasons and habitats (Wright 1995, Clarke 2000, Simpson and Norris 2000). The decision points chosen in Britain are the lower 5% of O:E values for ASPT and the lower 10% for number of taxa and BMWP (Wright 1995). These different cut-off levels have been chosen because variation in O:E values for number of taxa and BMWP score is greater than for ASPT, largely because the former indices are more dependent on sampling effort and less precisely predicted by the model (Wright 1995, Clarke 2000).

A modification of this approach has been adopted in Canada with the BEAST method, where individual sites are compared with a single group of reference sites to which they have been "matched" based on their environmental characteristics (Reynoldson et al. 1995, 2000). In this case, probability ellipses are constructed around the most similar group of reference sites. The community at the test sites is then compared directly with the community at the reference sites in ordination space. If the test site falls outside the 90% probability ellipse it is deemed to have failed because its community is considered different from those in reference sites with similar environmental characteristics (Reynoldson et al. 2000).

We have implemented this approach in the case studies below, by calculating the multivariate (Bray–Curtis) distance of each reference site's community to the center of its faunal group. If a test site community is predicted to be in a certain faunal group, and the multivariate distance of the test site's community to the center of its group is above the 90th percentile of distances for the reference sites, we deem that the test site has failed. The BEAST approach as described corresponds to a Type I error probability of 10%.

The decision points described use the 25th, 10th or 5th percentiles of the Reference Condition distribution, depending on variability in the biota descriptor; that is, the ecological meaning that might be attributed to a given degree of deviation from Reference Condition. Only the 5th percentile for the British ASPT is as stringent as significance levels commonly used for statistical hypothesis tests ($\alpha = 0.05$). While "$\alpha = 0.05$" is widely used as the necessary evidence of "real" deviation from the null hypothesis in science, it is based more on convention than explicit scientific or management considerations.

All of the forgoing discussion about decision points has only been based on the null hypothesis distribution and the desired Type I error probability; that is, the proportion of sites in Reference Condition that would be mistakenly failed with a particular decision point. Recently in general ecology, and particularly in biological assessment (e.g., Quinn and Keough 2002, Downes et al. 2002), there has been greater interest in the power of hypothesis tests (that is, the chance of failing a site that is indeed degraded), and its converse, the probability of Type II error (the chance of passing a site that is degraded). Sokal and Rohlf (1981) proposed a Type II error probability of 20% (80% power) in their worked example and this seems to have been adopted with little discussion, when Type II errors are considered at all in assessment studies (Norris and Georges. 1993).

We feel that the most important decision is the relative importance of Type I and Type II errors, and therefore that an analysis of their size relative to a given decision point is essential in any biological assessment. Determination of the Type II error probability requires a set of observations that are known to deviate from the null hypothesis by a certain *effect size* (Figure 5-3). This amounts to an *alternative hypothesis distribution*, and is by far the most difficult and controversial aspect of the calculations usually known as *power analysis*. Clearly, we cannot use observations from test sites that are only suspected of being degraded because the degree to which they are damaged (or even if they really are damaged) will be unknown.

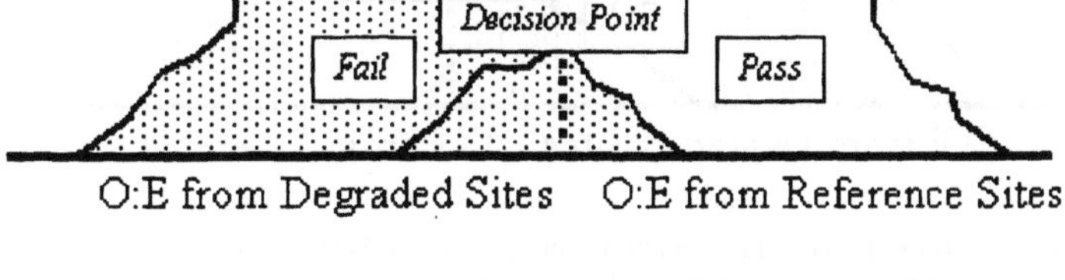

Figure 5-3. Given a pass–fail decision point based on variability in the Reference Condition (the null hypothesis distribution) and the desired Type I error probability, and a distribution of known degraded sites with a given effect size, we can calculate the proportion of degraded sites that pass (Type II error).

For our analysis of Type II error probabilities, we created a set of sites where the biota have a simulated, defined deviation from Reference Condition. Since these sites had a simulated impact on the biota, we called them "Simpacted" sites. The Simpacted sites started as the complete set of biota and environmental data from the reference sites. Then, the biota were modified in some way that we felt approximately corresponded to degradation from probable stressors. We have created Simpacted communities that indicate an impact of elimination of pollution-sensitive taxa (loss of taxonomic richness). Simulated impacts could also be created for cultural eutrophication, with reduction in abundance depending on the stressors known, or expected, in the region under consideration. The example presented with the case studies is the Diversity Simpact, where any Ephemeroptera, Plecoptera, or Trichoptera (EPT) taxon in a given community has a 50:50 chance of elimination. Once the reference biota is exposed to the Simpact, we take this Simpacted site and put it through the assessment exactly as if it was a test site. The proportion of Simpacted sites that pass the assessment represents the Type II error probability for that Simpact and decision point. In our examples, we use a decision point based on a Type I error probability of 10%, and note the resulting value for the Type II error probability. If we feel that the two types of error are equally important, or Type II errors are more important to avoid than Type I errors, we may then wish to move the decision point to adjust the error probabilities (Figure 5-4).

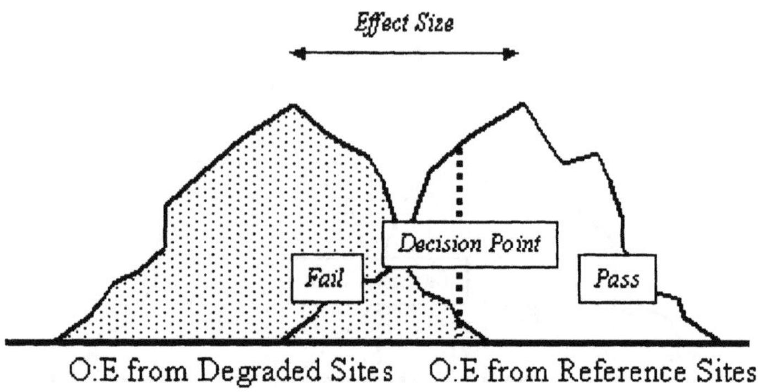

Figure 5-4. Adjusting the pass–fail decision point (cf Figure 5-3), making it easier to fail a site and therefore more likely that a site in Reference Condition will be mistakenly failed (Type I error) and less likely that a degraded site will be mistakenly passed (Type II error).

Clearly the definition of the Simpacted communities is key in assessing Type II error. It is based on ecological and management notions of the magnitude of deviation from Reference Condition it is important to detect. If the Simpact has a very small, ecologically insignificant, effect on the biota, the effect size will be small, and the corresponding Type II error probability will be very large (Figure 5-5). Some may feel that a better Diversity Simpact would be elimination of *all* (rather than just a randomly chosen 50%) of the sensitive taxa. Our point is not to argue for a particular degree or nature of simulated degradation, but to encourage the scientists carrying out the assessment to have the discussion and consider Type II error quantitatively.

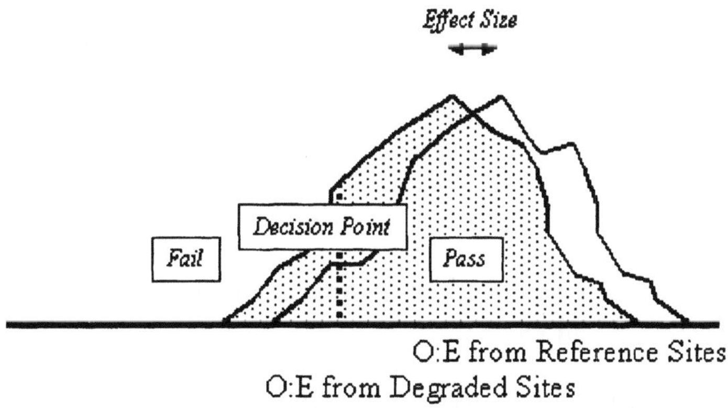

Figure 5-5. A small difference between the known degraded sites and the reference sites will translate into a small effect size and a large Type II error probability (cf Figure 5-3).

3. SENSITIVITY OF ASSESSMENTS: HOW OFTEN DO REAL TEST SITES FAIL?

In addition to the Simpacted sites, we also present the results of test site assessment for each of the case studies. This is an opportunity to compare several descriptors of the biota in terms of their sensitivity to deviation from Reference Condition, and to quantify the degree of agreement among the different descriptors in their assessments.

The test sites are actual sites that did not fulfill the criteria of being in Reference Condition because they are exposed to putative stressor(s). It should be noted that, other than knowing that they are exposed, we do not know *a priori* if these sites are actually environmentally damaged. We use the decision point(s) arrived at by consideration of Type I and Type II errors to decide whether a given test site should pass or fail. In our case studies, this decision point is just that which separates the most extreme 10% of values in the null hypothesis from the 90% of less extreme values. The proportion of test sites failing with a given descriptor is one measure of the sensitivity of assessment with that descriptor; that is, the greater the proportion failing, the more sensitive the assessment.

4. CASE STUDIES: DECISION-MAKING

In the case studies, we look at the Type II error probabilities and the sensitivity of assessments with O:E ratios of taxa (AUSRIVAS approach; Simpson and Norris 2000), multivariate distance of a test community from the center of its predicted group (BEAST approach; Reynoldson et al. 1995), and a multiple regression approach with the deviation between the expected and observed multivariate distance to the median reference community (MCDist approach).

4.1 Near shore Environment of North American Great Lakes

Not surprisingly, in the Great Lakes study, the Diversity Simpact (50% of EPT taxa eliminated from each reference site to simulate an impacted test site) did not differ much from the reference sites with respect to descriptors of the biota (Table 5-1). There were not many Ephemeroptera, Plecoptera, or Trichoptera in the Great Lakes benthic communities in Reference Condition, so the Simpact did not affect the communities much. Therefore, in assessing Type II error with this Simpact, we bear in mind that we are effectively looking at our ability to detect very small effect size (Figure 5-5). The test sites, as a group, were not notably lower in Taxon Richness than the reference sites, and the average multivariate distance between a site and the centre of its predicted group was about the same in the Simpacted and test sites when compared to the reference sites.

Assessments were done of the Diversity Simpact and test sites, using the AUSRIVAS O:E50, BEAST, and MCDist approaches. Using a Type I error probability of 10% (i.e., a pass–fail decision point at the 90th percentile of BEAST and MCDist residuals and a 10th percentile decision point for AUSRIVAS O:E scores), there was considerable variation among the three assessment approaches in results from both the Simpacted and actual test communities.

The percentage of Simpacted sites that pass is a measure of the Type II error probability. With all three assessment approaches, the Type II error rate was very high: 75–90% (Table 5-2). This is really an indication of the inappropriateness of this particular Simpact for the Great Lakes study, as noted above, and indicates the need to carefully design the Simpact to suit the particular assessment context. There was some lack of consistency among the three approaches in their assessment of actual test sites.

The AUSRIVAS approach failed the smallest percentage of test sites (7%) MCDist residuals based on multiple regression and the BEAST approaches were more sensitive (42–46% of test sites failed). It appears that the test sites in the Great Lakes study deviate from Reference Condition more in proportional composition, which is considered by the BEAST and MCDist approaches, and less in the actual occurrence of taxa, which is the primary consideration in the AUSRIVAS approach.

Table 5-1. Mean (standard deviation) of descriptors of biota at reference, Diversity Simpact, and test sites in the Great Lakes study.

	Taxon richness	MCDist	AUSRIVAS O:E50	BEAST
Reference (n=252)	7.6 (2.8)	0.59 (0.15)	1.01 (0.18)	0.35 (0.15)
Diversity Simpact (n=252)	7.5 (2.7)	0.58 (0.09)	1.01 (0.18)	0.35 (0.12)
Test (n=57)	7.8 (4.0)	0.59 (0.06)	0.97 (0.17)	0.43 (0.14)

Table 5-2. Assessment of Diversity Simpacted and test sites from the Great Lakes study using AUSRIVAS, BEAST, and MCDist approaches. Failed sites were lower than the 10th percentile of AUSRIVAS O:E50 values and higher than the 90th percentile of BEAST and MCDist values from the reference site distribution (see text).

	Diversity Simpacted sites n=252		*Test sites* n=57	
	Pass	*Fail*	*Pass*	*Fail*
AUSRIVAS	229 (91%)	23	53	4 (7%)
BEAST	212 (84%)	40	33	24 (42%)
MCDist	189 (75%)	63	31	26 (46%)

4.2 Fraser River Biomonitoring Program

For this assessment of streams in the Fraser River basin, with EPT taxa much more evident than in the Great Lakes study, biota in the sites where the Diversity Simpact was applied differed markedly from the reference biota (Table 5-3). As in the Great Lakes study, the test sites as a group were similar in Taxon Richness to the reference sites, but MCDist was smaller in test sites, indicating less heterogeneity among the test sites relative to the reference sites. Both the mean MCDist and the O:E50 values of the Diversity Simpacted sites were quite different from the mean of the reference sites, indicating that this Diversity Simpact had a strong effect on the biota.

Assessments of the Diversity Simpact and test sites, again using Type I error rate of 10% as a decision criterion, showed much lower Type II error rates (8–16%), with all three assessment methods, than in the Great Lakes study (Table 5-4). This is really an indication of the greater impact of the Diversity Simpact on the communities (i.e., a bigger effect size), thus making their deviation from reference more detectable than in the Great Lakes study.

In assessment of the actual test sites, the BEAST approach was most sensitive to deviations from Reference Condition, failing more than half of the test sites (Table 5-4). The residuals of MCDist from a multiple regression only resulted in failure of 17% of the test sites. The structure of communities in the Fraser River reference sites differed mainly in presence of taxa, relative to test sites, making deviations from Reference Condition more detectable with either the AUSRIVAS or BEAST approaches.

5-3. Mean (standard deviation) of descriptors of biota at reference, Diversity Simpact, and test sites in the Fraser River Study.

	Taxon richness	MCDist	AUSRIVAS O:E50	BEAST
Reference (n=219)	15.3 (4.0)	0.61 (0.17)	1.02 (0.20)	0.36 (0.11)
Diversity Simpact (n=219)	10.2 (3.1)	0.62 (0.09)	0.61 (0.17)	0.60 (0.11)
Test (n=57)	15.1 (3.1)	0.47 (0.07)	0.89 (0.27)	0.40 (0.10)

Table 5-4. Assessment of Diversity Simpacted and test sites from the Fraser River Basin Study using AUSRIVAS, BEAST, and MCDist approaches. Failed sites were lower than the 10th percentile of AUSRIVAS O:E50 values and higher than the 90th percentile of BEAST and MCDist values from the reference site distribution (see text).

	Diversity Simpacted sites n=219		*Test site* n=23	
	Pass	*Fail*	*Pass*	*Fail*
AUSRIVAS	36 (16%)	183	12	11 (48%)
BEAST	17 (8%)	202	10	13 (57%)
MCDist	31 (14%)	188	19	4 (17%)

4.3 Spring and Autumn Assessment of Streams in the Australian Capital Territory

In the Australian Capital Territory study, there were some interesting patterns in the effect of the Diversity Simpact on Autumn versus Spring reference biota (Table 5-5). There was a noticeably bigger difference between the mean of the reference sites and both the Diversity Simpacted and test sites for Taxon Richness, MCDist, AUSRIVAS O:E50, and BEAST biota descriptors.

This pattern is confirmed in the assessments of both Simpacted and actual test sites (Table 5-6). The Type II error rates were in the range 19–44% in the Autumn assessment, based on the proportion of sites subjected to the Diversity Simpact that passed the assessment. This compares favorably to the Spring assessment of the Diversity Simpact, which had Type II error probabilities of 32–50%.

For actual test sites, from 58–71% of the sites sampled in Autumn failed, while only 43–60% of Spring sites failed. Clearly the sites are more sensitively assessed in Autumn than in Spring. All three approaches were fairly sensitive to deviations from Reference Condition in the test sites, indicating effects of stressors on both presence or absence and proportional composition on test communities.

Table 5-5. Mean (standard deviation) of descriptors of biota at reference, Diversity Simpact, and test sites in the Australian Capital Territory study.

	Taxon richness	MCDist	AUSRIVAS O:E50	BEAST
Reference				
Autumn (n=93)	18.6 (3.6)	0.38 (0.07)	1.01 (0.14)	0.35 (0.07)
Spring (n=88)	17.3 (3.2)	0.34 (0.06)	1.00 (0.13)	0.33 (0.06)
Diversity Simpact				
Autumn (n=93)	15.0 (3.2)	0.48 (0.07)	0.80 (0.14)	0.49 (0.08)
Spring (n=88)	14.2 (2.7)	0.39 (0.07)	0.84 (0.14)	0.41 (0.08)
Test				
Autumn (n=24)	12.7 (4.4)	0.50 (0.09)	0.74 (0.18)	0.48 (0.11)
Spring (n=30)	13.9 (5.2)	0.39 (0.10)	0.86 (0.26)	0.42 (0.12)

Table 5-6. Assessment of Diversity Simpacted and test sites from the Australian Capital Territory Study using AUSRIVAS, BEAST, and MCDist approaches. Failed sites were lower than the 10th percentile of AUSRIVAS O:E50 values and higher than the 90th percentile of BEAST and MCDist values from the reference site distribution (see text).

	Diversity Simpacted sites Autumn n=93 Spring n=88		*Test sites* Autumn n=24 Spring n=30	
	Pass	*Fail*	*Pass*	*Fail*
AUSRIVAS	41 (44%)	52	8	16 (67%)
	44 (50%)	44	17	13 (43%)
BEAST	18 (19%)	75	7	17 (71%)
	28 (32%)	60	12	18 (60%)
MCDist	29 (31%)	64	10	14 (58%)
	42 (48%)	46	16	14 (47%)

This chapter has primarily dealt with making decisions with different assessment approaches using the RCA. Different approaches may result in conflicting conclusions of pass and fail. For example, in the UK, three different assessments are provided for each site. With nearly 30,000 assessments done throughout the country in 1990, standardized rules are needed to manage possible differences. The final class adopted is the median of the three outputs, except where the band for ASPT is lower, in which case it is adopted (Wright 1995). The issue was also considered in Australia and there it was decided to use the lowest of the taxa or SIGNAL index (Chessman 1995) or individual season or habitat assessments (Simpson and Norris 2000). This latter approach clearly is based on taking a precautionary position for environmental protection.

One example from our case studies illustrates these sometimes similar and sometimes mixed messages in assessments. In assessment of the actual test sites, the AUSRIVAS and BEAST approaches were quite consistent; there were only two sites passed by AUSRIVAS that BEAST failed, and only one site failed by AUSRIVAS that BEAST passed (Table 5-7). However, it was quite a different story when the Diversity Simpacted sites were assessed. Over a third of the sites failed by the BEAST approach were passed by AUSRIVAS. Inconsistencies like this, whether detected for actual or Simpacted sets of sites, indicate which assessment approaches are best suited for detecting particular types of impacts.

Table 5-7. Consistency in assessment of test sites in Autumn sample of Australian Capital Territory streams

		AUSRIVAS		
		Pass	*Fail*	*TOTAL*
	Pass	6	1	7
BEAST	*Fail*	2	15	17
	TOTAL	8	16	24

Table 5-8. Inconsistency in assessment of Diversity Simpact sites in Autumn sample of Australian Capital Territory streams

		AUSRIVAS		
		Pass	*Fail*	*TOTAL*
	Pass	14	4	18
BEAST	*Fail*	27	48	75
	TOTAL	41	52	93

In the next chapter we describe how assessments with apparently conflicting outcomes can be combined into an overall assessment of a site, and show how combining assessment decisions with other information can move us beyond a simple assessment decision to an hypothesis about the causes of deviation from Reference Condition at a site.

5. POINTS TO REMEMBER

— Passing or failing sites is a test of the null hypothesis that a test site is in Reference Condition. Uncertainty in making a decision about the site is dependent on how uncertain we are about the characterization of the distribution of the Reference Condition, and the biota and their environment at the test site.

— Establishment of the probability of mistakenly failing (Type I error) or passing (Type II error) a test site will be guided by knowing the variability of biotic descriptors of the Reference Condition and the relative importance of the two types of errors in terms of management responses and costs.

— The distribution of a biological descriptor among the reference sites enables the setting of the Type I error pass–fail boundary. Boundaries have often been set quite conservatively at levels up to 25%.

— Determining the likelihood of a Type II error in the Reference Condition Approach requires a set of observations that are *known* to deviate from the Reference Condition. These observations may be from experimentally manipulated sites, or from reference site data manipulated in known ways. They cannot be from sites suspected of being damaged but whose actual status is unknown.

Chapter 6

BEYOND PASS AND FAIL

Interpreting and Using Assessments

In the previous chapters we have described the Reference Condition Approach for assessing freshwater ecosystems. The method has several novel aspects that are advances in aquatic ecosystem assessment. It moves beyond more traditional assessment methods by describing variation in relatively widespread reference sites, by developing predictive models that relate biota to the characteristics of their environment, and by setting appropriate decision-making criteria that explicitly take uncertainty into account. However, just as for traditional methods, the end result is a determination that a site is or is not in Reference Condition. For management decisions this is still only a first step.

In this chapter we outline a post-decision analysis intended to take biological assessment from a tentative judgment of pass or fail, and extend it to look further and hypothesize about the cause of the detected impacts. We illustrate this post-decision analysis with data from four test sites in streams in the Australian Capital Territory. The sites were sampled in both Autumn and Spring (Nichols et al. 2000) as part of a long-term monitoring program, several years after the data discussed in previous chapters.

1. SEVERITY OF FAIL: MAGNITUDE AND NATURE OF DEVIATION FROM REFERENCE CONDITION

The degree to which a test site deviates from Reference Condition can be an important component in defining a post-decision strategy. In Chapter 5 we discussed statistical and management issues that define whether a test site "passes" or "fails" relative to the Reference Condition. Clearly, some failed sites will be in worse condition than others, and a reasonable question often asked when a site fails is "How bad is it?". The question can be partly answered by looking at the size of the site's deviation from Reference Condition.

If a site fails, the actual degree of deviation from Reference Condition ("severity of fail") can give some indication of the severity of damage. Also, the damage can be ranked against a series of categories of damage (known as "bands"), which simplify graphical or map presentation of the assessments and aid interpretation. Bands are an integral part of many assessment methods, such as RIVPACS (Clarke 2000) and AUSRIVAS (Simpson and Norris 2000), the BEAST (Reynoldson et al. 2000) and the IBI (Kerans and Karr 1994). The bands are often represented as different colors on maps; for example, see the Australian national State of the Environment Report (Ball et al. 2001).

Bands must be defined sensibly or else they can complicate matters. If bands are not carefully defined, sites that are actually similar to each other can appear on opposite sides of a band boundary and be treated to quite different management responses as a result. The number of categories that it is sensible to create will depend largely on two things; first, the variability of the index of deviation from Reference Condition (e.g. O:E value) in the reference sites and second, the variability of the index as estimated at one site (Clarke 2000). The smaller these variabilities are, the more confidence we will have that the assessment lies in a particular band and that it is a true assessment at the level reported. An example of how bands are derived for the O:E ratio from the AUSRIVAS approach to assessment (see Chapters 4 and 5) is shown in Table 6-1. Their interpretation is defined in Ball et al. (2001).

Table 6-1. Division of AUSRIVAS indices into bands for reporting (Ball et. al. 2001). O:E band values from one ACT model to illustrate width and variability of the Reference Condition (A band)

Band	O:E Values	Label	Interpretation
X	>1.17	**More biologically diverse than reference.** O:E greater than 90% of reference site O:E.	More families found than expected. Potential biodiversity "hot-spot". Mild organic enrichment. Continuous irrigation flow in a normally intermittent stream.
A	0.83 – 1.17	**Similar to Reference.** O:E within range of central 80% of reference site O:E.	Expected number of families within the range found at 80% of the reference sites
B	0.49 – 0.82	**Significantly impaired.** O:E below 10th percentile of reference sites. Same width as Band A.	Fewer families than expected. Potential impact on water and/or habitat resulting in a loss of families.
C	0.15 – 0.48	**Severely impaired.** O:E below Band B. Same width as Band A.	Many fewer families than expected. Loss of families from substantial impairment of expected biota caused by water and/or habitat quality.
D	0 – 0.15	**Extremely impaired.** O:E below Band C to 0.	Few of the expected families remain. Severe impairment.

Using these bands, we categorize the results for four test sites from the Australian Capital Territory (Table 6-2). The table shows that site 20 is in Reference Condition, and its biota either have not suffered damage or they have possibly been affected by mild nutrient enrichment. The interpretation is based both on the site's current O:E score of 1.11 and on its previous spring assessment when it was in Band X. Sites 196 and 242 are in poor condition because their O:E indices are all well down in the B band, and Site 189 is even poorer. These assessments are also quite consistent through time, and that lends support to the conclusions from this assessment (see Hill 1965).

Table 6-2. AUSRIVAS assessments for four sites in the Australian Capital Territory in Autumn 2002, and Spring and Autumn 2001.

Site	O:E Taxa, Autumn 2002	Autumn 2002	Spring 2001	Autumn 2001
20	1.11	A	X	A
196	0.60	B	D	B
242	0.49	B	B	B
189	0.39	C	D	C

2. WHY DID IT FAIL?

2.1 Composition and ecology of the biota

Many programs estimate the proportional abundance of the taxa in the community at reference and test sites. We can use the data to make some interpretation based on our knowledge of the sorts of organisms and their relative abundance. In the ACT example (Figure 6-1), the large proportional abundances of Chironomidae and particularly Oligochaeta at sites 196 and 242 are consistent with our expectation if the site has little dissolved oxygen because of decomposition of organic material. At site 189 the composition suggests other stressors may be having an effect on the community. Interpretations such as these can only be made if we have some knowledge of the ecological characteristics of the fauna. Numerical dominance of these two groups has long been considered indicative of organically enriched conditions (Kolkwitz and Marsson 1909, Hynes 1960), and this has been confirmed by numerous studies.

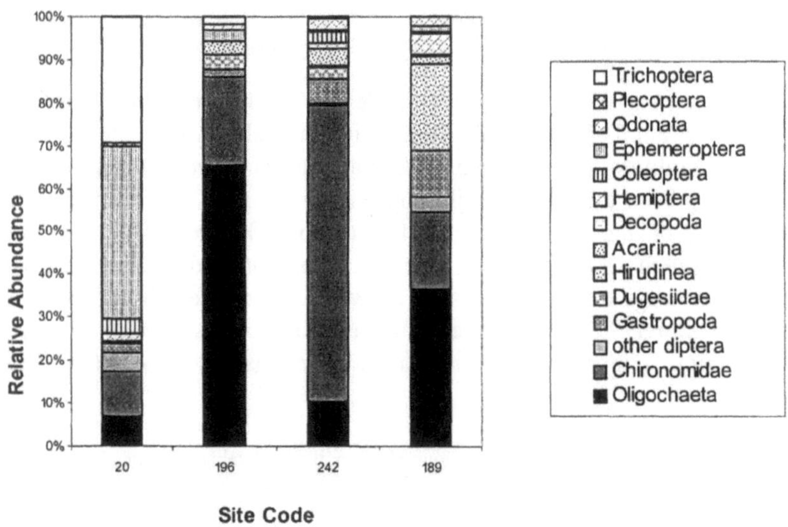

Figure 6-1. Relative abundance of selected macroinvertebrate taxa collected from four test sites in Autumn in the Australian Capital Territory.

"Pollution sensitivity" information for taxa in the community of interest has often been used to produce taxon-specific sensitivity or tolerance scores. Tolerance scores for Australian invertebrates have been incorporated into the SIGNAL biotic index (Chessman 1995). Similar indices are used in many places; for example in Britain (ASPT, Armitage et al. 1983), the USA (Hilsenhoff 1988) and New Zealand (MCI, Stark 1993). These and other biotic indices (see Norris and Georges 1993, Chapter 1) are based on knowledge of the ecological responses of invertebrates to mostly organic pollution. When we use such indices in assessment of lakes and rivers, we are really just formally applying our ecological knowledge of the responses of community members to pollution.

RIVPACS and AUSRIVAS models predict the taxa that will be present a site if it is in Reference Condition; therefore, we can also list the taxa expected but not found. Combined with the sensitivity scores of taxa, this information can enhance our diagnoses of the stressors most likely to be affecting a site. In Table 6-3, we can see that several sensitive taxa (SIGNAL score >5) were missing from sites 196, 242, 189. Other expected taxa that were absent at test sites were ones that are considered to be tolerant of pollution (low scores, Chessman 1995), so their absence suggests that stressors other than or as well as organic pollution may be imposing an effect. Thus, the fauna composition (including absent but expected taxa), combined with a knowledge of the sensitivity of the taxa to environmental stress, adds confidence to our conclusions from the initial assessment and suggests possible causes of damage.

Table 6-3. Pollution sensitivity scores (SIGNAL, Chessman, 1995) of taxa predicted to have a >50% chance of occurrence at four sites in the Australian Capital Territory. SIGNAL scores are from 1 (tolerant) to 10 (sensitive). "X" indicates the taxon was not found at the site. Shading indicates sites where biological condition was assessed as below Reference Condition according to O:E scores (Table 6-2).

Sites	**SIGNAL**	**20**	**196**	**242**	**189**
Band	**Score**	**A**	**B**	**B**	**C**
Missing Taxa					
Leptophlebiidae	10		X	X	X
Podonominae	9		X		
Gripopterygidae	9			X	X
Conoesucidae	9			X	X
Elmidae	8		X	X	X
Calmoceratidae	8		X		
Leptoceridae	8		X		X
Synlestidae	7				X
Gomphidae	6		X		
Ancylidae	5				
Ceratopogonidae	5		X		
Simuliidae	5		X		
Tanypodinae	5	X	X	X	X
Hydroptilidae	5			X	X
Ecnomidae	5			X	X
Sphaeriidae	4		X		
Hydrophilidae	4				
Caenidae	4		X		
Hydrobiidae	3				X
Planorbidae	3		X		
Corixidae	3				X

2.2 Habitat and water quality characteristics

Physical and chemical measures (Table 6-4) identified that two test sites (196, 242) of the four had potentially harmful low concentrations of dissolved oxygen, and two test sites (242, 189) had greater than desirable turbidity, compared to regional water-quality trigger values (ANZECC & ARMCANZ 2000). Site 189 also had relatively high conductivity levels that, while in themselves unlikely to be harmful, indicate concentrated inputs from some nearby sources. These data corroborate the interpretation that sites 196 and 242 were being affected by organic pollution, that site 196 is also being affected by sedimentation (high turbidity) and that there are some possible toxic effects at site 189. However, these were just spot measurements that may not be related to long-term conditions, which could be better, or worse. Other physical or chemical factors that were not measured could have damaged the biota. These are all reasons why biota, rather than only physical and chemical measurements, provide a useful measure of river condition. Chemical data can provide additional information that may support overall conclusions and indicate a source of impact for any sites that have failed to pass the decision point.

Table 6-4. Water quality characteristics of test and reference sites in the ACT region, Autumn 2002. Bold text indicates values outside recommended guideline levels (ANZECC & ARMCANZ 2000).

	Water temp (°C)	Conductivity (µS cm⁻1)	pH	Alkalinity (mg L^{-1} $CaCO_3$)	Dissolved Oxygen (mg L^{-1})	Turbidity (NTU)
Guideline levels						
Urban streams			**6.0–9**		**>6**	**<10**
Rural streams			**6.5–9**		**>4**	**<10**
Test Sites (O:E Band)						
20 (A)	**17.96**	**86.1**	**7.53**	**39**	**9.97**	**2.1**
196 (B)	**16.98**	**243**	**7.06**	**76**	3.84	**6.8**
242 (B)	**16.18**	**292**	**6.81**	**76**	2.47	11.8
189 (C)	**18.90**	**637**	**8.22**	**160**	**12.82**	28.7

2.3 Combining information from all sources

All the information about test sites discussed above can be summarized (Table 6-5) to inform and provide added confidence in the final assessment. The decision points for pass or fail were determined by checking the biota sampled at a test site against the list expected if the site were in Reference Condition. The comparison is just the first important step in determining the condition of a site. Here it is concluded that sites 189, 196 and 242 were clearly affected by several identifiable impacts, some of which were noted at the time of sampling. Together, this information supports our initial assessment on the biological condition or "health" of the sites, made by using the Reference Condition Approach, and it provides insights into the cause of their status that may be important for management.

The levels of evidence needed to support a final conclusion and subsequent management action for a set of test sites may vary depending on several factors. If there is a high environmental risk that, say, an endangered

Table 6-5 Combined assessment information for test sites in the Australian Capital Territory, including: the overall Autumn 2002 site assessment; comparisons with Spring and Autumn 2001 assessments shown in brackets; percentage contribution of numerically dominant taxa; number of predicted but missing taxa; potential impact indicators; and the overall conclusion. Site and habitat assessments are summarized from site information and habitat assessment at the time of sampling.

Site	Band (spring, Autumn)	Water quality characteristics	Numerically Dominant Taxa	# taxa predicted but not collected	Site and habitat assessment	Conclusion
20	**A (X,A)**	**Within guidelines**	**Ephemeroptera 40% Chironomidae 10%**	**1**	**Rural setting but riparian zone in good condition and fenced. No obvious stock access or run off from surrounding land.**	**Site is in good condition**
196	**B (D,B)**	**Low dissolved oxygen**	**Oligochaeta (67%), Chironomidae (20%)**	**12**	**Flecks of oil on the water surface, lots of sediment plume, silt deposit, anaerobic sediment odors; >90% of the edge habitat covered by periphyton and macrophytes; 35–65% of reach and edge covered by filamentous algae; 65–90% of the reach covered by periphyton, 30% of the of the substratum in the edge habitat and the reach was covered by detritus (sticks wood), Muck/mud (fine organics) covered 20% of the substratum, willow root mats; dominated by pool habitat; Habitat availability is low.**	**Suffers from urban run off and a small dam. Taxonomic composition indicates severe damage but taxa may be available for recolonization with rehabilitation**
242	**B (B,B)**	**Low dissolved oxygen, elevated turbidity**	**Chironomidae (71%)**	**10**	**Anaerobic sediment odors, some oil sheen observed on the water surface, turbid water, Much sediment plume, silt deposits, willows dominate the riparian zone, 35–65% of reach and edge covered in periphyton; 10–35% of edge covered by filamentous algae; 10–15% of the substratum covered by detritus (wood sticks) poor habitat, missing riffles/runs.**	**Site is significantly degraded and in need of rehabilitation**
189	**C (D,C)**	**Elevated turbidity and conductivity**	**Oligochaeta (45%), Chironomidae (23%)**	**11**	**Much sediment plume, sand and silt deposits, muck/mud (black, very fine organic), anaerobic sediment odor covering 30% of the substrate in the reach, periphyton and filamentous algae covering 90% of reach, upstream sections concreted and receives urban run off.**	**Site varies from extremely to severely degraded. Severe effects of urban run off. Site in need of rehabilitation**

species or a unique ecosystem may be lost if not protected, it may be appropriate to err on the side of lower levels of confidence that an effect has actually occurred. If the outcomes are likely to be less severe, or critical, or perhaps have major financial implications, more stringent tests may be appropriate. In either case, the levels of certainty that will be accepted should be proposed *a priori* (see Downes et al. 2002), and when met, the decision should be accepted.

3. SCENARIO BUILDING: WHAT WILL BE THE EFFECT OF REHABILITATION OR STRESS?

Scenario building, or the ability to answer "what if" questions, is a potentially useful, fundamental tool largely missing from aquatic bioassessment as it is currently executed. The Reference Condition Approach, as described in this book, has been widely used in aquatic biomonitoring for impact assessments (Wright 1995; Simpson and Norris 2000). Following our assessment, we can further analyze the data collected as part of the study, and use the results to suggest the cause of degradation, as described in the previous section. However, it is often desirable to predict what will happen to the biota, over a given time-scale, when test sites either are rehabilitated or they suffer increased exposure to stressors. Despite the range of prediction tools available, we are not aware of any in the aquatic literature that have been successfully developed for scenario building.

An extension of the Reference Condition Approach allows us to build scenarios that take into account the natural environmental features of a site. A generalized, admittedly speculative model (Figure 6-2) should accurately predict the possible effects of rehabilitation, or increased stressors. The biota are likely to respond differently to rehabilitation or to increased exposure to stress. There may be some threshold before degradation occurs and rehabilitation may be difficult because many associated features might be involved (Hobbs and Norton 1996).

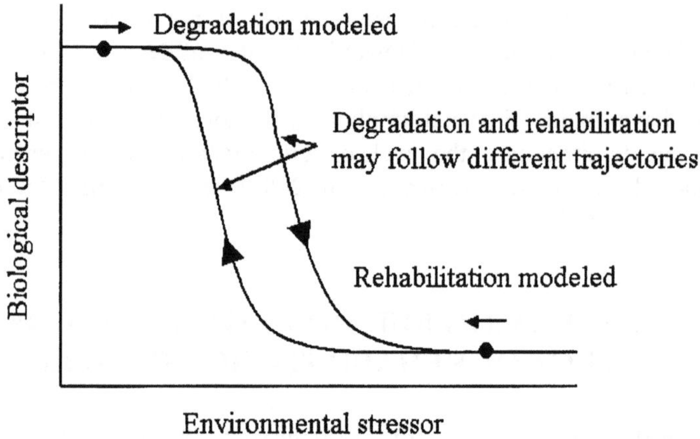

Figure 6-2. A generalized, speculative scenario for degradation of a site in good condition and rehabilitation of a degraded site. It is likely that the paths taken for each scenario will differ, with there being some resistance before a good site degrades and a lag in recovery for a poor site before associated factors also improve and a response is seen in the biota.

One of the strengths of the Reference Condition Approach is that it uses a large pool of sites to establish the Reference Condition, rather than one or two. Therefore, a potential application of models relating the biota to the environment, similar to those described in Chapter 4, would be to predict possible change resulting from management interventions for damaged areas, or the effects of proposed developments. Ecosystem restoration projects for example, would benefit from the ability to predict the likely biological consequences of actions before they are implemented. It would also be of significant benefit to be able to predict the effects of increased human activity on in-stream biota.

For example, we could build a scenario from relationships between predictor variables (say, flow and taxonomic richness in a simple linear model, Figure 6-3) modeled using the Reference Condition, and another variable that causes stress, such as concentration of nitrogen. Test sites that were not in Reference Condition because of high nitrogen concentrations could be compared to the scenario, and then we would be able to see the likely effects of reducing nitrogen concentration towards those occurring in natural conditions. In this simple example, we could estimate how much taxonomic richness would improve in a site with a given flow rate as nitrogen concentrations are reduced (Figure 6-3).

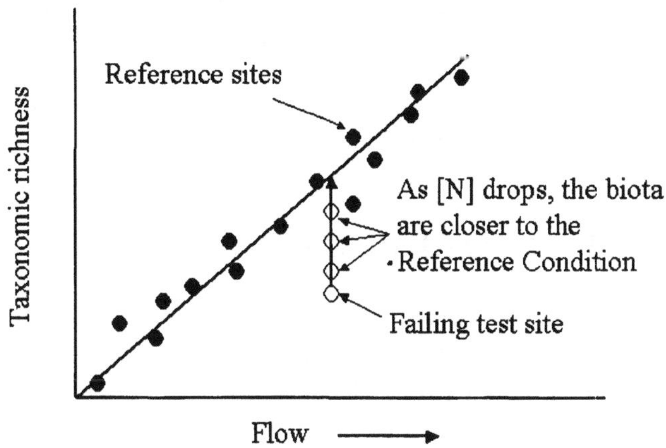

Figure 6-3. A linear model relating taxonomic richness to flow. The stressor variable, concentration of nitrogen, is subsequently associated with the Reference Condition. The likely taxonomic richness of the failing test site can be predicted as nitrogen concentration is changed to a level comparable to reference sites that have environmental characteristics similar to those of the failing site.

Armitage (2000) investigated the potential of using RIVPACS for predicting the effects of a specific stressor on the biota in a stream and he concluded that the addition of polluted sites to RIVPACS models would provide a possible tool for assessing pollution responses in a range of river types. In this case we would build a model based on test and reference sites together, which indicate variability in the nature and magnitude of the effects of a stressor on the biota. Unlike the assessment models presented earlier, no site assessment would be produced, because there would be no 'reference' condition as such for comparison. However, such models would provide a way to predict an expected taxonomic composition, or other descriptor, at a site of interest without the need for a large number of sites in Reference Condition. It could potentially be used in areas where the Reference Condition is unknown, or unavailable.

In summary, for assessments alone, in accounting for variation in the Reference Condition so that comparisons will be accurate and test sites can be matched with reference sites, the predictor variables must be those *unlikely to be affected* by human activity. In contrast, for scenario building, it is essential that the predictor variables *include* those responsible for damaging aquatic ecosystems and ones that may be managed. We identify the main driving influences on the biota, both natural and human, before restoration or development is undertaken. Such models would help us predict the likely biological outcomes of a management intervention that changed an environmental feature. Thus, we could assess the relative importance of factors controlling the biota, using models constructed this way. We could use the resulting information to assign priority to the protection or improvement of key environmental features.

4. POINTS TO REMEMBER

— Making a decision about whether a site is or is not in Reference Condition (i.e., passes or fails) is not the end of making an assessment.
— Severity of failing is an important piece of information that is useful in assessing a site. Bands of failure are commonly used to represent degrees of failure.
— Additional information such as composition and ecology of the biota, its tolerance or sensitivity to pollution, the habitat condition and water quality, are all likely to provide support for conclusions about site assessments, including the cause of any degradation observed.
— The Reference Condition Approach and associated modeling methods offer potential for building scenarios to evaluate the possible outcomes of rehabilitation or increased exposure to stressors.

Chapter 7

THE REFERENCE CONDITION APPROACH

In a nutshell

This book is a formal statement of the general principles and application of the Reference Condition Approach. So in this final chapter we draw together the major points of the book, describing the use of the Reference Condition Approach with benthic macroinvertebrate communities in the bioassessment of freshwater lakes and streams. We distil what we mean by the Reference Condition Approach, aspects of study design to identify the Reference Condition, how to explain and mathematically model variation, how to make decisions and use the outputs and, finally, where we see future developments taking place.

1. BIOASSESSMENT OF FRESHWATER ECOSYSTEMS USING THE REFERENCE CONDITION APPROACH

Freshwater ecological assessments collect information that is essential to freshwater management and protection. The Reference Condition Approach is a new paradigm in biological assessment, in which the "Reference Condition" is represented by many sites that are recognized to vary in their expression of good condition. The approach acknowledges that there is variability in biota and associated environmental characteristics among sites that are in acceptable (i.e. Reference) condition. Empirical modeling is used to explain as much of this natural variability as possible in relation to the environmental characteristics of the reference sites. The models predict which biota can be expected at a site being assessed if it is in Reference Condition. Deviation from Reference Condition is a measure of the effect of stressors on the ecosystem.

The Reference Condition Approach has evolved from biological assessment of still and running waters over the last century. The approach differs from traditional impact assessment by employing designs and analysis that avoid many of the statistical and resource constraints of more traditional methods. We believe that the Reference Condition Approach is a powerful tool that provides the first step in assessing the ecological health of a site. Once a site has been deemed to be in Reference Condition or not, further data or testing establishes the causes of the observed condition.

We consider a site to be "healthy" if it is in the condition we would expect to see if it were minimally exposure to the stressors of concern (i.e., if it is equivalent to Reference Condition). Health is defined relative to our expectations for particular systems rather than against an absolute benchmark of ecosystem attributes. The magnitude of a site's deviation from our expectations is a measure of how unhealthy its ecosystem is. This new paradigm of assessment is what we have come to call the Reference Condition Approach. Thus, the Reference Condition Approach provides an empirical definition of ecosystem health, although such a definition had previously proved elusive (e.g. Norris and Thoms 1999).

2. CASE STUDIES

The near shore, North American Great Lakes, the Fraser River (British Columbia) and the upper Murrumbidgee River (Australia) represent a variety of scales, water types and climates and have been used in this book as case studies to illustrate the Reference Condition Approach. These studies were chosen, in part, because they had common elements: collaborative objective setting, rigorous site selection criteria and standardized sampling and analysis. They demonstrate the general applicability and flexibility of the Reference Condition Approach.

3. DEFINING the REFERENCE CONDITION

Fundamental to the Reference Condition Approach is the identification and sampling of reference sites. The resulting database of biological and environmental attributes represents the Reference Condition.

Establishing the purpose of an assessment or monitoring study is a critical first step in the process of defining the Reference Condition. Unless all parties to the study have an explicit understanding of the reason for the study, it is almost inevitable that there will be some dissatisfaction with the outcome. To avoid dissatisfaction, all the parties involved in the study, including scientists and managers, should also be involved in defining the objectives.

The geographic and time boundaries and units of the study are influenced by geographic, physical or institutional factors, or some combination of these. When designing the study, there are several issues we must address before sampling commences, including the extent of the study, the types of habitats or water bodies to be sampled, the season of sampling, and the definition of an individual site. The grain size or site definition is particularly important because it determines many other study features such as sampling methods and the eventual number and location of sites. The Reference Condition Approach formally recognizes that the site, rather than points within it, is an individual "grain" in the study design and the scale of sampling replication.

Appropriate environmental descriptors of the sites should be considered at the study-design stage. They should include descriptors that can be measured at a variety of scales, from large (such as the catchment area upstream of a river site or of a lake) to small (such as bank-full width or sediment particle size). The features recorded for each site should include some that are unlikely to be related to human activities as well as some that will aid interpretation of possible deviation of test sites from the Reference Condition.

Criteria defining the reference state should be established well before the actual sampling, and rigorously applied. The structure of the macroinvertebrate community itself is not used to identify the site as reference or otherwise, thus avoiding any preconceived notions of the composition or diversity of the community at a "typical" reference site. Nor should there be any idea of the degree of variation of the community among the reference sites.

Sufficient reference sites should be sampled so that they encompass the range of variability in biota that has been minimally exposed to stressors of concern. The reference sites should cover the range of biota and habitat types likely to be encountered at test sites. It is the variation of the selected biotic descriptors among sites in Reference Condition, which is captured by sampling both the biota and their environments, that will describe the Reference Condition. The critical focus at the design stage is to adequately characterize the variation among communities within the geographic area and in habitat types that are likely to be represented at test sites. There is little sense in sampling reference sites (say, alpine streams) that have characteristics not represented among likely test sites (such as lowland rivers).

The descriptions of the biota and the environmental characteristics at a site form the basis of a predictive model for the Reference Condition. Subsequently they enable the expected biological condition at a test site to be determined, based on the environmental features at the site. For the benthic macroinvertebrate community, typically the data obtained will be the number of taxa at some predefined taxonomic level, abundance, or relative abundance (or biomass). These data may be modified before they are used in final analyses, by calculating integrated descriptors such as similarity indices, metrics or functional feeding groups. The format of descriptors of the biota will often be dependent on the method of collection. Regardless of the decisions concerning sampling methods and habitats, it is critical that the *sampling effort is equivalent at all sites,* reference and test. Many of the biological features of a site will be related to sampling effort and timing (e.g. seasonal effects), such as number of taxa, rare taxa, and abundance estimates. Ultimately, the Reference Condition Approach makes comparisons, and the confidence that can be placed in these partly will depend on ensuring that differences are unrelated to sampling.

The Reference Condition Approach involves large amounts of data, often from hundreds of sites, and requires carefully planned and managed data storage systems. The nature of the data, including biological and environmental records, is suited to some form of a relational database system. A database is much more efficient in both storage and data entry than common alternatives such as spreadsheets. Apart from sampling effort, intended comparisons made between the Reference Condition and test sites may be affected by inadvertent errors in the data. Confidence in assessments will be maintained if every aspect of a Reference Condition Approach bioassessment is automatically subjected to quality assurance and control.

4. DESCRIBING AND MODELING VARIATION IN THE REFERENCE CONDITION

The Reference Condition Approach requires that natural variability among the freshwater ecosystems present in the region of interest is adequately sampled, but only in places that match the types of sites that we intend to assess. Estimating and explaining variation in the Reference Condition is not a problem. Rather, the variation can be used to advantage and must be measured and characterized before assessment of a test site can be undertaken.

Once the data have been acquired, several decisions are needed on their treatment (including the elimination of rare taxa, presence/absence versus abundance, transformation) before proceeding with analyses. These decisions should be based on the ecological features that are seen as important for the study (based on the objectives), and characteristics of the study region that might predicate the importance of abundance measures ahead of presence/absence data.

The basic properties of the dataset should be examined as an important first step before more complex analysis. This step enables the context and boundaries of a study to be determined and provides information that will aid later interpretation. In addition to methods such as plots and correlations, this initial evaluation of the variation in Reference Condition is often done using classification and ordination, both multivariate methods. Classification is most useful for forming groups of sites with relatively homogeneous biota. Ordination is useful for evaluating the relationships among sites, and making graphical representation of groups formed by classification.

Once the we have described the variation among the environments and biota of the reference sites, we can model relationships between the biota and the environments. With the models we can predict the fauna we can expect to find at test sites that have similar environmental characteristics. In these models it is critical that relationships between the biota and the environment of reference sites are based only on environmental variables that are unrelated to potential stressor(s) of concern. Otherwise the model and prediction will be influenced by the stressor variable and provide an inappropriate prediction of the fauna in the absence of the stressor. However, no model will be perfect, and therefore it will be easier to evaluate the properties and performance of the models if we make an initial description of variation in the data.

Selection of methods for associating environmental features with variation in the Reference Condition will depend, in part, on the nature of the data and the manner in which they are treated. Usually there will be many environmental features that are to be associated with the biotic data. The biotic data may be in one of three forms. They may be compressed to a univariate response variable such as species richness or a diversity index; or the taxa may be used to classify sites into groups with relatively homogeneous biota; or the taxa information may be reduced to several multivariate descriptors. The first type of data has single biotic descriptors, either univariate or multivariate, that are most typically modeled with a multiple regression approach to establish predictive relationships. The second data type uses groups of sites based on their biota, and many environmental variables are associated with the groups, usually using discriminant function analysis to create the model. The third data type uses ordination of the sites based on their biotic composition (many biota), and associates many environmental variables with the ordination to produce the model. In each case, the models will predict the fauna likely to occur at test sites that are similar to Reference Condition in their environmental characteristics.

Even after application of a powerful predictive model, the inherent variability in environmental and biotic values of a set of reference sites means that deviations of reference sites from Reference Condition will form a residual distribution. The Reference Condition Approach assesses the values measured at any test site in comparison to the appropriate reference site distribution, testing the null hypothesis that there is no difference in the biota from Reference Condition. The goal is to decide whether or not the test site values could be part of the reference site distribution. The more the variation explained, the more sensitive the model to testing deviations from the Reference Condition. Reference sites that fall near the boundaries of the model distribution should be reviewed, but only removed if it can be clearly demonstrated that the data have suffered from poor sampling (over or under), or some previously unknown stressor. However, even the best models will have some unexplained, or residual, variation and this should be considered in terms of the model quality and the accuracy of predictions possible from the models. The residual variation is the uncertainty or imprecision in the application of our predictive model to a test site to generate an expectation of the biota at the test site if it is in Reference Condition. The less residual variation we have among reference sites, the more sensitive our assessment of test sites will be, and the smaller the deviation of the test site's biota from the expected Reference Condition we will be able to detect when making a decision on the condition of the test site.

5. DECISION-MAKING

The standard scientific approach to hypothesis testing is to establish the null hypothesis, collect data, and determine the probability that those data would have been collected if the null hypothesis were true. Uncertainty in making a decision about test site(s) in the Reference Condition Approach is determined by our uncertainty about the distribution of the Reference Condition, as well as the biota and their environment at the test site.

Decisions about the relative importance of mistakenly labeling a test site poor (Type I error) or in Reference Condition (Type II error) will be guided by knowing the variability of biotic descriptors of the Reference Condition. Other guiding influences will be the responses and costs associated with the assessment and the relative importance of each type of error. Regardless of importance accorded to each type of error, it is crucial that we know the risks involved if the errors occur. If environmental protection is a primary concern, we should err on the side of failing sites that may not be damaged. This approach can be justified because the consequences of damage can be serious and costly. The converse may be true if costly mitigation is envisaged. In such cases we may want to err on the side of passing sites that may be different if there is a high public cost to taking action, or in balancing where mitigation resources are to be used.

We must evaluate the likelihood of mistakenly failing a test site relative to the variation in the biological descriptors we have modeled for the Reference Condition. The likelihood of mistakenly passing a test site is more difficult to evaluate, and requires a set of observations that are known to deviate from the Reference Condition. These observations may be from experimentally manipulated sites, or from reference site data manipulated in known ways, as we have demonstrated above [give section number]. Such data cannot be from sites suspected to be suffering the effects of a stressor but whose status is unknown. The decision on the level of confidence we wish to adopt in relation to mistakenly passing a test site will be dependent on the variability of the biological descriptor of the Reference Condition, and the size of the change (effect size) that it is desirable to detect. Small differences will be difficult to detect and will often result in a high likelihood of mistakenly passing test sites.

6. BEYOND PASS AND FAIL

The Reference Condition Approach moves beyond traditional assessment methods, and involves describing variation in natural communities, developing predictive models for assessing aquatic ecosystem health, and setting appropriate decision-making criteria. While these are all advances in aquatic ecosystem assessment, the end result is still a determination of whether or not a site is or is not in Reference Condition: that is, it passes or fails. However, for management decisions, this is still a first step and much more can be done with this approach.

The degree to which a test site deviates from Reference Condition, or the "severity of fail", may be an important component of post-decision assessment. Categories or 'bands' indicating the degree of damage can simplify graphical or map presentation of the assessments and aid interpretation. However, banding needs to be used with caution because it is a simplification of assessments that actually lie on a continuum and with some level of uncertainty associated with them.

After the initial assessment is made, additional information such as composition and ecology of the biota, pollution tolerance or sensitivity, habitat condition and water quality are all likely to provide support for the conclusions about site assessments. All these sources of information should be compiled formally to both strengthen the final conclusion and indicate possible causes of impact on the biota.

The Reference Condition Approach offers the potential to build scenarios that take into account the natural environmental features of a site. The Reference Condition can be associated with natural levels of agents known to cause stress in the region of interest. Once this association has been identified, the levels of these stressors found at poor-condition test sites can be modeled through their natural ranges to provide scenarios of likely improvement for a given change in the stressor. Alternatively the large number of reference and test sites can be combined, and variables known to cause stress, which are explicitly excluded from the standard Reference Condition Approach, can be used to explain variation in the whole data set. Such models could be used to predict possible responses in the biota where the stressors are improved or degraded as intended if various management options were implemented.

7. FINISH

In this book we have described both what the Reference Condition Approach is and how it differs from more traditional assessment methods; how one would actually go about implementing an Reference Condition Approach bioassessment program; and some of the issues involved in acquiring the necessary information. Using three case studies, we have demonstrated how predictive models are built, and the strengths and weaknesses of different approaches for assessing sites. Finally we have speculated on where the Reference Condition Approach may lead in the future. Whether or not we have been successful we will leave the reader to judge.

References

Abel, P.D. (1989). *Water pollution biology.* (Ellis Horwood Ltd., Chichester, England.) 231pp.

Allan, J.D. (1984). Hypothesis testing in ecological studies of aquatic insects, pp. 484-507. In: V.H. Resh and D.M. Rosenberg (eds.). The ecology of aquatic insects. Praeger Publishers, NY. 625 pp.

Allan, J.D., D.L. Erickson, and J. Fay. (1997). The influence of catchment land use on stream integrity across multiple spatial scales. *Freshwater Biology* **37,** 149-161.

Anderson, M.J. (2001). A new method for non-parametric multivariate analysis of variance. *Austral Ecology* **26,** 32-46.

ANZECC & ARMCANZ (2000). Australian and New Zealand Guidelines for fresh and marine water quality –October 2000. (Australian and New Zealand Environment and Conservation Council and Resource Management council of Australia and New Zealand, Canberra.)

Armitage, P.D. (2000). The potential of RIVPACS for predicting the effects of environmental change. Pages 93-111 in J.F. Wright, D.W. Sutcliffe, and M.T. Furse (editors). Assessing the biological quality of fresh waters: RIVPACS and other techniques. Freshwater Biological Association, Ambleside, UK.

Armitage, P.D., Moss, D., Wright, J.F. and Furse, M.T. (1983). The performance of a new biological water quality score system based on macroinvertebrates over a wide range of unpolluted running-water sites. *Water Research* **17,** 333-347.

B.C. Ministry of Environment, Lands and Parks, and Environment Canada. (1991). State of the environment report for British Columbia. B.C. Ministry of Environment, Lands and Parks and Environment Canada, Victoria, B.C.

Bailey, R.C., Kennedy, M.G., Dervish, M.Z. and Taylor, R.M. (1998). Biological assessment of freshwater ecosystems using a reference condition approach: comparing predicted and actual benthic invertebrate communities in Yukon streams. *Freshwater Biology* **39** 765-774.

Bailey, R.C., Norris, R.H., Reynoldson, T.B. and Day, K.E. (1995). Macroinvertebrate community structure and sediment bioassay results from nearshore areas of North American Great Lakes. *Journal of Great Lakes Research.* **21**, 42-52.

Ball, J., Donnelley, L., Erlanger, P., Evans, R., Kollmorgen, A., Neal, B. and Shirley, M. (2001). Inland Waters, Australia State of the Environment Report 2001 (Theme Report), CSIRO Publishing on behalf of the Department of the Environment and Heritage Canberra. http://www.ea.gov.au/soe/2001/index.html.

Barbour, M. T. and J. Gerritsen, (1996). Subsampling of benthic samples: a defense of the fixed-count method. *Journal of the North American Benthological Society.* **15,** 386–391.

Barbour, M.T., J. Gerritsen, B.D. Snyder, and J.B. Stribling. (1999). Rapid Bioassessment Protocols for Use in Streams and Wadeable Rivers: Periphyton, Benthic Macroinvertebrates and Fish, Second Edition. EPA 841-B-99-002. (U.S. Environmental Protection Agency; Office of Water; Washington, D.C.)

Beak, T.W. (1965). A biotic index of polluted streams and its relationship to fisheries, pp.191-210. In: O. Jaag (ed.). *Advances in water pollution research.* Proceedings of the Second International Conference, Tokyo, August, 1964. Vol. 10 (Pergamon Press Ltd., Oxford, England.) 373 pp.

Belbin, L. (1987). PATN: pattern analysis package reference manual. Division of Wildlife and Rangelands Research, Commonwealth Scientific and Industrial Research Organization, Canberra, Australia. 337 pp.

Belbin, L., and C. McDonald. (1993). Comparing three classification strategies for use in ecology. Journal of Vegetation Science 4:341-348.

Bloom, S.A. 1981. Similarity indices in community studies: potential pitfalls. *Marine Ecology Progress Series*. **5,** 125-128.

Bray, J.R. and J.T. Curtis. (1957). Ordination of the upland forest communities of southern Wisconsin. *Ecological Monographs..* **27,** 325-349.

Brierley, G., K.F. Fryirs, and T. Cohen. (1996). Geomorphology and river ecology in southeastern Australia: An approach to catchment characterisation. Part one: A geomorphic approach to catchment characterisation. Working paper 9603, Graduate School of the Environment, Macquarie University, Sydney, Australia.

Brinkhurst, R.O. (1966). The Tubificidae (Oligochaeta) of polluted water. *Verhandlungen der Internationalen Vereinigung für Theoretische und Angewandte Limnologie*. **16,** 854-859.

Brookes A & Shields F.D. Jr (Eds) (1996) *River Channel Restoration: Guiding Principles for Sustainable Projects*. Wiley, Chichester. 433 p.

Brydges, T., and Lumb, A. (1998). Canada's Ecological monitoring and assessment network: where we are at and where we are going. *Environmental Monitoring and Assessment* **51**, 595-603.

Cairns, J. Jr. and W.H. Van Der Schalie. (1980). Biological monitoring. Part I - Early warning systems. *Water Research* **14**, 1179-96.

Canton, S.P. and J.W. Chadwick. (1988). Variability in benthic invertebrate density estimates from stream samples. *Journal of Freshwater Ecology* **4**, 291-7.

Cao, Y. (1999). Rare species are important in bioassessment. (Reply to the comment by Marchant). *Limnology and Oceanography*. **44**, 1841-1842.

Cao, Y., Larsen, D.P. and Thorne, R.St-J. (2001). Rare species in multivariate analysis for bioassessment: some considerations. *Journal of the North American Benthological Society*. **20**, 144-153.

Cao, Y., Williams, D.D., and Williams, N.E. (1998). How important are rare species in aquatic community ecology and bioassessment? *Limnology and Oceanography*. **43**, 1403-1409.

Carter, J.L. and Resh, V.H. (2001). After site selection and before data analysis: sampling, sorting, and laboratory procedures used in stream benthic macroinvertebrate monitoring programs by USA state agencies. *Journal of the North American Benthological Society*. **20**, 658-682.

Chapman, P.M. (1990). The sediment quality triad approach to determining pollution-induced degradation. *Science of the Total Environment*. **97/98**, 815-825

Chapman, P.M. Sediment quality criteria from the sediment quality triad: An example. (1986). *Environmental Toxicology and Chemistry*. **5**, 957-964.

Chessman, B. C. (1995). Rapid river assessment using macroinvertebrates: a procedure based on habitat specific sampling, family level identification and a biotic index. *Australian Journal of Ecology* **20**, 122-129.

Chessman, B.C. (1995). Rapid assessment of rovers using macroinvertebrates: A procedure based on habitat-specific sampling, family level identification and a biotic index. *Australian Journal of Ecology*, **20**, 122-129

Chutter, F.M. (1972). An empirical biotic index of the quality of water in South African streams and rivers. *Water Res*earch. **6**, 19-30.

Clarke, K. R. & Ainsworth, M. (1993) A method of linking multivariate community structure to environmental variables. *Marine Ecology Progress Series*, **92**, 205-219.

Clarke, K.R. and Warwick, R.M. (1994). *Change in Marine Communities: An Approach to Statistical Analysis and Interpretation.* Plymouth Marine Laboratory, Plymouth, U.K.

Clarke, R. (2000). Uncertainty in estimates of biological quality based on RIVPACS. Pages 39-54 in J.F. Wright, D.W. Sutcliffe, and M.T. Furse (editors). Assessing the biological quality of fresh waters: RIVPACS and other techniques. Freshwater Biological Association, Ambleside, UK.

Cochran, W.G. (1977). Sampling techniques. 3rd edition. (John WIley and Sons, New York). 428pp.

Cook, S.E.K. (1976). Quest for an index of community structure sensitive to water pollution. *Environ. Pollut*ion. **11,** 269-288.

Corkum, L.D. (1991). Spatial patterns of macroinvertebrate distributions along rivers in eastern deciduous forest and grassland biomes. *Journal of The North American Benthological Society*. **10**,358-371.

Corkum, L.D. (1992). Spatial distributional patterns of macroinvertebrates along rivers within and among biomes. *Hydrobiologia* **239,** 101-111.

Courtemanch, D. L., (1996). Commentary on the sub-sampling procedure used for rapid bioassessments. *Journal of the North American Benthological Society*. **15,** 381–385.

Cullen, P. (1990). Biomonitoring and environmental management. *Environmental Monitoring and Assessment* **14,** 107-114.

Cummins, K.W. and Klug, M.J. (1979). Feeding ecology of stream invertebrates. *Annual Review of Ecology and Systematics*. **10**, 147-172.

Davies, P. E. (Ed.) (1994). *Monitoring River Health Initiative. River Bioassessment Manual. National River Processes and Management Program*.(Freshwater Systems: Tasmania.)

Davies, P.E. (2000). Development of a national river bioassessment system (AUSRIVAS) in Australia. Pages 113-124, in J.F. Wright, D.W. Sutcliffe, and M.T. Furse (editors). Assessing the biological quality of fresh waters: RIVPACS and other techniques. Freshwater Biological Association, Ambleside, UK.

Davis W.S. & Simon T. (Eds) (1995) *Biological Assessment Criteria: Tools for Water Resource Planning and Decision-making*. Lewis Publishers, Boca Raton, Florida.

De Pauw, N.J. , Verreth, J. and Talloen, M. (1979). Biological assessment methods applied on the rivers Torrente parma, Torrente Strirone and Fiume Po, 33-93. In: *3rd Tech. Sem. Biological Water Assessment Methods, Parma October 1978. Vol 1.* (Ghetti, P.F. Ed.) Published for Commission of the European Communities.

Diamond J.M., Barbour M.T, and Stribling J.B. (1996). Characterizing and comparing bioassessment methods and their results: A perspective. *Journal of the North American Benthological Society*. **15,** 713-727

Dolédec, S. and Chessel, D. (1994). Co-inertia analysis: an alternative method for studying species-environment relationships. *Freshwater Biology* **31,** 277-294.

Downes, B.J., Barmuta, L.A., Fairweather, P.G., Faith, D.P., Keough, M.J., Lake, P.S., Mapstone, B.D. and Quinn, G.P. (2002). *Monitoring Ecological Impacts: Concepts and Practice in Flowing Waters*. (Cambridge University Press) 446 pp.

Downing, J.A. (1979). Aggregation, transformation, and the design of benthos sampling programs. *Journal of the Fisheries Board of Canada* **36,** 1454-63.

Dynesius, M. and Nilson, C. (1994) Fragmentation and flow regulation of river systems in the northern 3rd of the world. *Science*.**266,** 753-762.

Eberhardt, L.L, (1978) Appraising variability in population studies. *Journal of Wildlife Management*. **42,** 207-238

Eberhardt, L.L. (1976) Quantitative ecology and impact assessment. *Journal of Environmental Management*. **4**, 27-70.

Eberhardt, L.L., and Thomas, J.M.. (1991). Designing environmental field studies. *Ecological Monographs*. **61**, 53-73.

Elliott, J.M. 1977. *Some Methods for the Statistical Analysis of Samples of Benthic Invertebrates*. 2nd ed. Freshwater Biological Association Scientific Publication No. 25. 156 pp.

Environment Canada. (1995). *Measuring the health of the river. Environmental quality program 1995 status report*. (Environment Canada, North Vancouver, B.C. ;Fraser River Action Plan) 28pp.

Environment Canada. (1998). *Fraser River Action Plan, pollution abatement technical summary report*. (Environment Canada, North Vancouver, B.C.;Fraser River Action Plan) DOE FRAP 1998-01.

Faith, D.P. and Norris, R.H. (1989). Correlation of environmental variables with pattern of distribution and abundance of common and rare freshwater macroinvertebrates *Biological Conservation*. **50**, 77-98.

Faith, D.P. Minchin, P.R. and Belbin, L. (1987). Compositional dissimilarity as a robust measure of ecological distance. *Vegetatio* **69**, 57-68.

Faith, D.P., C.L. Humphrey and P.L. Dostine. 1991. Statistical power and BACI designs in biological monitoring: Comparative evaluation of measures of community dissimilarity based on benthic macroinvertebrate communities in Rockhole Mine Creek, Northern Territory, Australia. *Australian Journal of Marine and Freshwater Research*.. **42,** 589-602.

Feio, M.J., Graca, M, and Reynoldson, T.B. (2002). Development of an assessment and biomonitoring method for the water quality of the Mondego Basin. Spanish Limnology Association XI Congress and Iberian Limnology III Congress.

T.B.Frissell, C.A., L.J. Liss, C.E. Warren, and M.D. Hurley. (1986). A hierarchical framework for stream habitat classification: viewing streams in a watershed context. Environmental Management 10:199-214.

Furse, M.T., Moss, D., Wright, J.F. and Armitage, P.D. (1984). The influence of seasonal and taxonomic factors on the ordination and classification of running-water sites in Great Britain and on the prediction of their macro-invertebrate communities. *Freshwater Biology*. **14**, 257-280.

Gauch, H.G., Jr. (1982). Multivariate analysis in community ecology. Cambridge University Press, Cambridge, England 298 pp.

Gerritsen, J. (1995). Additive biological indices for resource management. *Journal of the North American Benthological Society*. **14**, 451-457.

Gerritsen, J., Barbour, M.T. and King, K. (2000). Apples, oranges, and ecoregions: on determining pattern in aquatic assemblages. *Journal of the North American Benthological Society*, **19**, 487-496.

Goodman, D. (1975). The theory of diversity-stability relationships in ecology. *Quarterly Review of Biology* **50,** 237-66.

Gore J.A. (Ed.) (1985) *The Restoration of Rivers and Streams*. Butterworth, Stoneham. 280 p.

Gray, C and T. Tuominen. (1998). Health of the Fraser River Aquatic Ecosystem Vol. 1. A synthesis of research conducted under the Fraser River Action Plan. DOE FRAP 1991-11, Environment Canada, Vancouver B.C.

Green, R.H. (1971). A multivariate statistical approach to the Hutchinsonian niche: bivalve molluscs of central Canada. *Ecology* **52,** 543-556.

Green, R.H. (1974). Multivariate niche analysis with temporally varying environmental factors. *Ecology* **55**, 73-83

Green, R.H. (1979). Sampling design and statistical methods for environmental biologists. John Wiley and Sons, New York, NY 257 pp.

Harper, D., and M. Everard. (1998). Why should the habitat-level approach underpin holistic river survey and management? *Aquatic Conservation: Marine and Freshwater Ecosystems* **8**, 395-413.

Hawkes, H.A. (1979). Invertebrates as indicators of river water quality. *In Biological Indicators of Water Quality.* Eds. A. James and L. Evison, Chap. 2 John Wiley Chichester, England.)

Hawkins, C.P. and Norris, R.H. (2000a). Effects of taxonomic resolution and use of subsets of the fauna on the performance of RIVPACS-type models. In: Assessing the biological quality of freshwaters: RIVPACS and similar techniques. Pp 217-228. Edited by J F Wright, D W Sutcliffe & M T Furse. Freshwater Biological Association, Ambleside.

Hawkins, C.P., Norris, R.H., Hogue, J.N. and Feminella, J.W. (2000b). Development and evaluation of predictive models for measuring the biological integrity of streams. *Ecological Applications* **10** (5) 1456-1477.

Hawkins, C.P., R.H. Norris, J. Gerritsen, R.M. Hughes, S.K. Jackson, R.K. Johnson, and R.J. Stevenson. (2000a). Evaluation of the use of landscape classifications for the prediction of freshwater biota: synthesis and recommendations. *Journal of The North American Benthological Society* **19**, 541-556.

Hellawell, J.M. 1986. Biological indicators of freshwater pollution and environmental management. (Elsevier Applied Science Publishers, London, England.) 546 pp.

Hernandez, J.M. (1997). *Database design for mere mortals: A hands-on guide to relational database design.* Addison-Wesley Developers Press.

Hill, A.B. (1965). The environment and disease: association or causation? *Proceedings of the Royal Society of Medicine* **58**, 295-300.

Hilsenhoff, W.L. (1977). *Use of Arthropods to evaluate water quality of streams.* Technical Bulletin 100. Dept Natural Resources, Madison WI.

Hilsenhoff, W.L. (1988). Rapid field assessment of organic pollution with a family-level biotic index. *Journal of the North American Benthological Society,* **7**, 65-78.

Hobbs, R.J. and Norton, D.A. (1996). Towards a conceptual framework for restoration ecology. *Restoration Ecology*, **4** 93-110.

Hogg, I.D. and Norris, R.H. (1991). Effects of runoff from land clearing and urban development on the distribution and abundance of macroinvertebrates in pool areas of a river. *Australian Journal of Marine and Freshwater Research.* **42**, 507-18.

Holdway, D.A., Brennan, S.E. and Ahokas, J.T. (1995). Short review of selected fish biomarkers of xenobiotic exposure with an example using fish mixed-function oxidase. *Australian Journal of Ecology* **20**, 34-44.

Howmiller, R.P. and Scott, M.A. (1977). An environmental index based on relative abundance of oligochaete species. *Journal of the Water Pollution Control Federation* **49,** 809-815.

Hughes R.M. & Larsen D.P. (1988). Ecoregions: an approach to surface water protection. *Journal of the Water Pollution Control Federation*, **60**, 486–493.

Hughes R.M. (1995) Defining acceptable conditions. *Biological Assessment Criteria: Tools for Water Resource Planning and Decision-making* (Eds

W.S. Davis & T. Simon), pp. 31–47. Lewis Publishers, Boca Raton, Florida.

Hughes R.M., Larsen D.P. & Omernik J.M. (1986) Regional reference sites: a method for assessing stream potentials. *Environmental Management*, **10**, 629–635.

Hughes, R.M., D.P. Larsen, and J.M. Omernik. 1986. Regional reference sites: a method for assessing stream potentials. Environmental Management 10:629-635.

Humphrey, C.L. Storey, A.W. and Thurtell, L. (2000). AUSRIVAS: operator sample processing errors and temporal variability – implications for model sensitivity. Pages 143-163 in J.F. Wright, D.W. Sutcliffe, and M.T. Furse (editors). Assessing the biological quality of fresh waters: RIVPACS and other techniques. Freshwater Biological Association, Ambleside, UK.

Hunsaker, C.T. and Levine, D.A. (1995). Hierarchical approaches to the study of water quality in rivers. *Bioscience*. **45,** 193-203.

Hurlbert, S.H. 1971. The nonconcept of species diversity: a critique and alternative parameters. *Ecology* **52,** 577-86.

Hurlbert, S.H. 1984. Pseudoreplication and the design of ecological field experiments. *Ecological Monographs* **54,** 187-211.

Hutchinson, G.E. (1968). When are species necessary? In: Lewontin E. (ed.), *Population Biology and Evolution* (Syracuse University Press, Syracuse, N.Y.) pp 177-186.

Hynes H.B.N. (1975) Edgardo Baldi Memorial Lecture. The stream and its valley. *Verhandlungen der Internationalen Vereinigung für Theoretische und Angewandte Limnologie*, **19**, 1–15.

Hynes, H.B.N. (1993). Historical Perspective and future direction of biological monitoring of aquatic systems. Pp. 11-22, in S.L. Loeb and A. Spacie (eds.) *Biological Monitoring of Aquatic Systems*. (Lewis Publishers, Boca Raton, Florida, USA.)

Hynes, H.B.N. 1960. *The Biology of Polluted Waters*. (Liverpool University Press, Liverpool, England.) 202 pp.

International Joint Commission. 1987. Report on Great Lakes Water Quality. Report to the International Joint Commission, Great Lakes Water Quality Board. Windsor, Ontario. 236pp.

International Joint Commission. 1988. Procedures for the assessment of contaminated sediment problems in the Great Lakes. Report to the Great Lakes Water Quality Board. Windsor, Ontario. 140 p.

International Organisation of Standardisation (ISO). (1979). *Assessment of the biological quality of rivers by a macroinvertebrate "score"*. ISO/TC 147/SC 5/WG 6 N 5: 18pp.

Jackson, D.A. 1993. Multivariate analysis of benthic invertebrate communities: the implication of choosing particular data standardizations, measures of association, and ordination methods. *Hydrobiologia* **268,** 9-26.

Johnson, R. K. (1998). Spatio-temporal variability of temperate lake macroinvertebrate communities: detection of impact. *Ecological Applications* **8**:61–70.

Johnson, R.K. and Goedkoop, W. (2000). The use of biogeographical regions for partitioning variance of littoral macroinvertebrate communities. *Verhandlungen der Internationalen Vereinigung für theoretische und angewandte Limnologie* **27**, 333-339.

Karr J.R. (1991) Biological integrity: a long-neglected aspect of water resource management. *Ecological Applications*, **1**, 66–84.

Karr, J. R., and Chu, E. W. (1998). Restoring life in running waters: better biological monitoring. Island Press, Covelo, California.

Karr, J.R. (1991) Biological integrity: a long-neglected aspect of water resource management. *Ecological Applications*, **1**, 66–84.

Karr, J.R. (1999). Defining and measuring river health. *Freshwater Biology* **41**,

Karr, J.R. 1981. Assessment of biotic integrity using fish communities. *Fisheries* **6**, 21-27.

Karr, J.R. and Chu, E.W. (1999). *Restoring Life in Running Waters. Better Biological Monitoring*. (Island Press, Washington, DC.) 206 pp.

Karr, J.R. and Chu, E.W. (2000). Sustaining living rivers. *Hydrobiologia* **422/423**, 1-14.

Keough and Mapstone (1997). Designing environmental monitoring for pulp mills in Australia . *Water Science and Technology*. **35**, 397-404.

Kerans, B. L. & J. R. Karr, 1994. A benthic index of biotic integrity (B-IBI) for rivers of the Tennessee Valley. *Ecological Applications* **4**: 768–785.

King, D.L. and Ball, R.C. (1964). A quantitative biological measure of stream pollution. *Journal of the Water Pollution Control Federation*. **36**, 650-653.

King, D.L. and Ball, R.C. 1964. A quantitative biological measure of stream pollution. *Journal of the Water Pollution Control Federation*. **36**, 650-653.

Klemm, D.J., Blocksom, K.A., Thoeny, W.T., Fulk, F.A., Herlihy, A.T. Kaufmann, P.R. and Cormier, S.M. (2002). Methods development and use of macroinvertebrates as indicators of ecological conditions for streams in the Mid-Atlantic Highlands Region. *Environmental Monitoring and Assessment*. **78**, 169-212.

Knight T.W. & Morris D.W. (1996) How many habitats do landscapes contain? *Ecology*, **77**, 1756–1764.

Kolkwitz, R. and Marsson, M. 1909. Okologie der tierischen Saprobien. Beitraga zur Lehre von des biologischen Gewasserbeurteilung. *Internationale Revue der gesamten Hydrobiologie und Hydrographie* **2**, 126-152.

Lake, P.S. and Barmuta, L.A. (1986). Stream benthic communities: Persistent presumptions and current speculations. In; P. De Deckker and W.D. Williams (eds) '*Limnology in Australia*' (CSIRO. Dr. W. Junk).

Legendre, P. and Legendre, L. (1998). *Numerical Ecology*. 2nd English edition, (Elsevier Science, Amsterdam.)

Lemly, A.D. (1997). Role of season in aquatic hazard assessment. *Environmental Monitoring and Assessment*. **45**, 89-98.

Linke, S., Bailey, R.C. and Schwindt, J. (1999). Temporal variability of stream bioassessments using benthic macroinvertebrates. *Freshwater Biology*. **42**, 575-584.

Long, E.R. and P.M. Chapman. 1985. A sediment quality triad: measures of sediment contamination, toxicity and infaunal community composition in Puget Sound. Mar. Pollut. Bull. 16: 405-415.

Mapstone, B. D. (1995). Scalable decision rules for environmental impact studies: effect size, type I and type II errors. Ecological Applications 5:401–410. planning and decision making. Lewis Publishers, Boca Raton, Florida.
Marchant, R. (1999). How important are rare species in aquatic community ecology and bioassessment? A comment on the conclusions of Cao et al. *Limnology and Oceanography*. **44**, 1840-1841.
Marchant, R. 1989. A subsampler for samples of benthic invertebrates. *Bulletin of the Australian Society of Limnology* **12** 49--52.
Marchant, R., A. Hirst, R. H. Norris, R. Butcher, L. Metzeling, and D. Tiller. 1997. Classification and prediction of macroinvertebrate assemblages from running waters in Victoria, Australia. Journal of the North American Benthological Society 16:664--681.
Marchant, R., Barmuta, L. and Chessman, B. (1995). A preliminary study of the ordination and classification of macroinvertebrate communities from running waters in Victoria, Australia. *Australian Journal of Marine and Freshwater Research* **45**, 945-962
Marchant, R., Mitchell, P. and Norris, R. (1984). The distribution of benthic invertebrates along a disturbed section of the LaTrobe River, Victoria. An analysis based on numerical classification. *Australian Journal of Marine and Freshwater Research.* **35**, 355-74.
Marchant, R., P. Mitchell and R.H. Norris. 1984. Distribution of benthic invertebrates along a disturbed section of the LaTrobe River, Victoria: An analysis based on numerical classification. *Australian Journal of Marine and Freshwater Research...* **35**, 355-374.
McCune, B. and Mefford, J.J. (1999). PC-ORD. Multivariate Analysis of Ecological data, Version 4. (MjM Software Design, Gleneden Beach, Oregon, USA.)
McDonald, M.E. (2000). EMAP overview: objectives, approaches, and achievements. *Environmental Monitoring and Assessment* **64**, 3-8.
Merritt & Cummins 199#
Merritt, R.W. and Cummins, K.W. (eds) (1984). *An Introduction to the Aquatic Insects of North America*, 2nd ed. (Kendall/Hunt Publishing,
Metcalfe, J. L. (1989) Biological water quality assessments of running waters based on macroinvertebrate communities: history and present status in Europe. *Environmental Pollution*. **60**, 101-139.
Milbrink, G. (1973). Commuhities of Oligochaeta as indicators of the water quality in Lake Hjalmaren. *Zoon* **1**, 77-88.
Minshall, G.W. (1984). Aquatic insect-substratum relationships. *The Ecology of Aquatic Insects* (eds V.H. Resh and D.M. Rosenberg), pp. 358-400 (Praeger, New York).
Morin, A. (1985). Variability of density estimates and the optimization of sampling programs for stream benthos. Canadian Journal of Fisheries and Aquatic Sciences 42: 1530-40.
Morris S.E. (1995) Geomorphic aspects of stream-channel restoration. *Physical Geography*, **16**, 444–459.
Moss, D., M.T. Furse, J.F. Wright, and P.D. Armitage. (1987). The prediction of the macro-invertebrate fauna of unpolluted running-water sites in Great Britain using environmental data. Freshwater Biology 17: 41-52.

Needham, P.R. and R.L. Usinger. 1956. Variability in the macrofauna of a single riffle in Prosser Creek, California, as indicated by the Surber sampler. *Hilgardia* **24**, 383-409.
Newall, P. and Wells, F. (2000). Potential for delineating indicator-defined regions for streams in Victoria, Australia. *Journal of the North American Benthological Society*. **19**, 557-571.
Newbury, R.W. 1984. Hydrologic determinants of aquatic insect habitats. Pages 323-357 in V.H. Resh and D.M. Rosenberg (eds) The ecology of aquatic insects. Praeger, New York.
Nichols, S., Sloane, P., Coysh, J., Williams, C., and Norris, R. (2000). *Australian Capital Territory, AUStralian RIVer Assessment System.* (Cooperative Research Centre for Freshwater Ecology, University of Canberra ACT 2601.)
Norris, R.H. (1986). Mine waste pollution of the Molonglo River, New South Wales and the Australian Capital Territory. Effectiveness of remedial works at Captain's Flat mining area. *Australian Journal of Marine and Freshwater Research* **37**, 147-57.
Norris, R.H. (1995). Biological monitoring: the dilemma of data analysis. *Journal of the North American Benthological Society.* **14,** 44-450.
Norris, R.H. (1996). Predicting water quality using reference conditions and associated communities. In *Study Design and Data Analysis in Benthic Macroinvertebrate Assessments of Freshwater Ecosystems using a Reference Site Approach.* Eds. R.C. Bailey, R.H. Norris and T.B. Reynoldson. Pp. 32-52. (Technical Information Workshop. North American Benthological Society. 44th Annual Meeting, Kalispell, Montana.
Norris, R.H. and Georges, A. (1986). Design and analysis for assessment of water quality. In: *Australian Limnology* (Eds. De Decker P. and Williams, W.D.) (CSIRO, Melbourne).
Norris, R.H. and Georges, A. (1993). Analysis and interpretation of benthic surveys. In: *Freshwater Biomonitoring and Benthic Macroinvertebrates* (Eds. D.M. Rosenberg and V.H. Resh) (Chapman and Hall.) 488 pp.
Norris, R.H. and Hawkins, C.P. (2000). Monitoring river Health. *Hydrobiologia* **435**, 5-17
Norris, R.H. and Norris, K.R. (1995). The need for biological assessment of water quality: Australian perspective. *Australian Journal of Ecology* **20**, 1-6.
Norris, R.H. and Thoms, M. (1999) What is River Health? *Freshwater Biology* **41**, 197-210.
Norris, R.H., Lake, P.S. and Swain, R. (1982). Ecological effects of mine effluents on the South Esk River, North East Tasmania. III. Benthic macroinvertebrates. *Australian Journal of Marine and Freshwater Research.* **33**, 789-809.
Norris, R.H., McElravy, P. and Resh, V.H. (1992). The sampling problem. In: P. Calow and G.E. Petts (Eds.) *Rivers Handbook*: Volume 1 (Blackwell Scientific Publications, Oxford England.) 526 pp.
Norris, R.H., Prosser, I., Young, B., Liston, P., Bauer, N., Dyer, F., Linke, S., Thoms, M., (2001) The assessment of River Condition (ARC). An audit of the ecological condition of Australian Rivers. (Environment Australia) (http://audit.ea.gov.au/ANRA/water/docs/river_assessment.pdf)

Odum, E.P., (1971). *Fundamentals of Ecology* 3rd Edition. (W.B. Saunders Company, Philadelphia.)
Omernik J.M. (1995) Ecoregions: a spatial framework for environmental management. *Biological Assessment Criteria: Tools for Water Resource Planning and Decision-making* (Eds W.S. Davis & T. Simon), pp. 49–62. Lewis Publishers, Boca Raton, Florida.
Omernik, J. M. (1995). Ecoregions: a spatial framework for environmental management. Pages 49–62 *in* W. S. Davis and T. P. Simon (editors). Biological assessment and criteria. Tools for water resource planning and decision making. Lewis Publishers, Boca Raton, Florida.
Omernik, J. M., and Griffith. G. E. (1991). Ecological regions versus hydrologic units: frameworks for managing water quality. Journal of Soil and Water Conservation 46:334–340.
Omernik, J.M. (1987). Ecoregions of the conterminous United States. *Annals of the Association of American Geographers* **77**, 118-125
Ontario Ministry of Environment. 1990. The Canadian Great Lakes basin intake outfall atlas (8 vol.) Ed. M. Griffiths. Water Resources Branch, Toronto, Ontario.
Ormerod, S.J. and R.W. Edwards. (1987). The ordination and classification of macroinvertebrate assemblages in the catchment of the River Wye in relation to environmental factors. Freshwater Biology 17: 533-46.
Parsons M. & Norris R.H. (1996) The effect of habitat-specific sampling on biological assessment of water quality using a predictive model. *Freshwater Biology*, **36**, 419–434.
Patrick, R. (1949). Diatom communities. In: *The Structure and Function of Freshwater Microbial Communities*, (ed. J. Cairns, Jr).. Research Division Monograph 3, Virginia Polytechnic Institute and State University, Blacksburg, VA. pp. 151-64
Peters, R.H. (1991). *A critique for ecology* (Cambridge University Press).
Petts G.E. (1989) Historical analysis of fluvial hydrosystems. *Historical Change of Large Alluvial Rivers: Western Europe* (Eds G.E. Petts, H. Moller & A.L. Roux), pp.1–18. Wiley, Chichester.
Pielou, E.C. (1984). The interpretation of ecological data. John Wiley and Sons, NY. 263 pp.
Plafkin J.L., Barbour M.T., Porter K.D., Gross S.K. & Hughes R.M. (1989) *Rapid Bioassessment Protocols for Use in Streams and Rivers: Benthic Macroinvertebrates and Fish*. United States Environmental Protection Agency, EPA/440/4-89/001, Washington, DC.
Poff, N.L. (1997). Landscape filters and species traits: towards mechanistic understanding and prediction in stream ecology. *Journal of the North American Benthological Society*. **16**, 391-409.
Policansky, D. (1998). Science and decision making for water resources. Ecological Applications 8, 610-618.
Pontasch, K.W., E.P. Smith and J. Cairns Jr. 1989. Diversity indices, community comparison indices and canonical discriminant analysis: Interpreting the results of multispecies toxicity tests. *Water Research* **22**,1229-1238
Quinn, G.P. and Keough, M.J. (2002). *Experimental Design and Data Analysis for Biologists*. (Cambridge University Press, Cambridge UK.) 537 pp.

Rabeni, C.F. and Wang, N. (2001). Bioassessment f streams using macroinvertebrates: are the Chironomidae necessary? *Environmental Monitoring and Assessment.* **71**, 177-185.

Rabeni, C.F., N. Wang, and R.J. Sarver. (1999). Evaluating adequacy of the representative stream reach used in invertebrate monitoring programs. *Journal of The North American Benthological Society* **18**, 284-291.

Rapport D.J. (1989) What constitutes ecosystem health? *Perspectives in Biology and Medicine*, **33**, 120–132.

Reese, P.F., Reynoldson, T.B., Richardson, J.S., and Rosenberg, D.M. (2001). Implications of seasonal variation for biomonitoring with predictive models in the Fraser River catchment, British Columbia. Can. J. Fish Aq. Sci. 58: 1411-1418.

Resh, V.H. (1979). Sampling variability and life history features: basic considerations in the design of aquatic insect studies. Journal of the Fisheries Research Board of Canada. 36: 290-311.

Resh, V.H. and McElravy, E.P. (1993). Contemporary quantitative approaches to biomonitoring using benthic macroinvertebrates. Pp 159-94, in D.M. Rosenberg and V.H. Resh (eds.) *Freshwater Biomonitoring and Benthic Macroinvertebrates* (Chapman and Hall, New York.)

Resh, V.H., Norris, R.H. and Barbour, M.T. (1995). Design and implementation of rapid assessment approaches for water resource monitoring using benthic macroinvertebrates. *Australian Journal of Ecology.* **20**, 108-121.

Reynoldson, T.B and Wright, J.F. (2000) The reference condition: problems and solutions. Pages 293-304, in *Assessing the biological quality of fresh waters: RIVPACS and other techniques*. (Eds. J.F. Wright, D.W. Sutcliffe, and M.T. Furse). Freshwater Biological Association, Ambleside, UK.

Reynoldson, T.B. (1948). An ecological study of the Enchytraeid worm population of sewage bacteria beds: Synthesis of field and laboratory data. *Journal of. Animal Ecology* **17,** 27-38.

Reynoldson, T.B., Bailey, R.C., Day, K.E. and Norris, R.H. (1995). Biological guidelines for freshwater sediment based on BEnthic Assessment of SedimenT (the BEAST) using a multivariate approach for predicting biological state. Australian Journal of Ecology. **20**:198-219.

Reynoldson, T.B., Day, K.E. and Pascoe, T. (2000). The development of the BEAST: a predictive approach for assessing sediment quality in the North American Great Lakes. Pages 165-194 in J.F. Wright, D.W. Sutcliffe, and M.T. Furse (editors). Assessing the biological quality of fresh waters: RIVPACS and other techniques. Freshwater Biological Association, Ambleside, UK.

Reynoldson, T.B., Norris, R.H., Resh, V.H., Day, K.E. and Rosenberg, D.M. (1997). The reference condition: a comparison of multimetric and multivariate approaches to assess water-quality impairment using benthic macroinvertebrates. *Journal of the North American Benthological Society.* **16**, 833-452.

Richards, C., Johnson, L.B. and Host, G.E. (1996). Landscape-scale influences on stream habitats and biota. Canadian Journal of Fisheries and Aquatic Sciences 53 (Supplement 1):295–311.

Rodriguez, P. and Wright, J.F. (1991). Description and evaluation of a sampling strategy for macroinvertebrate communities in Basque rivers (Spain). Hydrobiologia. **213,** 113-124.

Rosenberg, D.M, Reynoldson, T.B., and Resh, V.H. (1998). Establishing reference conditions in the Fraser River catchment, British Columbia, Canada. FRAP Rep. No. DOE-FRAP 1998-32. Fraser River Action Plan, Environment Canada, Vancouver B.C.

Rosenberg, D.M. and Resh, V.H. (1993). *Freshwater Biomonitoring and Benthic Invertebrates.* (Chapman and Hall, New York.) 488 pp.

Scrimgeour G.J. & Wicklum D. (1996) Aquatic ecosystem health and integrity: problems and potential solutions. *Journal of the North American Benthological Society.* **15**, 2154–2261.

Sheldon, F., and K.F. Walker. (1998). Spatial distribution of littoral invertebrates in the lower Murray-Darling River system, Australia. Marine and Freshwater Research 49:171-182.

Shelford, V.E. (1911). Ecological succession: stream fishes and the method of physiographic analysis. *Biological Bulletin* (Woods Hole) **21**, 9-34.

Silberschatz, A., Korth, H.F., and Sudarshan, S. (1997). *Database System Concepts.* 3rd Ed. McGraw-Hill Companies, Inc.

Simpson, E.H. (1949). Measurement of diversity. *Nature* (London) 163: 688.

Simpson, J.C., and R.H. Norris. (2000). Biological assessment of river quality: development of AUSRIVAS models and outputs. Pages 125-142 in J.F. Wright, D.W. Sutcliffe, and M.T. Furse (editors). Assessing the biological quality of fresh waters: RIVPACS and other techniques. Freshwater Biological Association, Ambleside, UK.

Sokal, R.R. and F.J. Rohlf. 1981. Biometry. 2nd ed. W.H. Freeman and Company, London, England. 859 pp.

Southwood, T.R.E. (1977). Habitat, the templet for ecological strategies? Journal of Animal Ecology 46:337-365.

Sovell, L.A. and Vondracek, B. (1999). Evaluation of the fixed-count method for rapid bioassessment Protocol III with benthic macroinvertebrate metrics. *Journal of the North American Benthological Society.* **18**, 420-426

Stark, J.D. (1993). Performance of the Macroinvertebrate Community Index: effects of sampling method, sample replication, water depth, current velocity, and substratum on index values. *New Zealand Journal of Marine and Freshwater Research,* **27**, 463-478.

Statzner, B, and Higler, B. (1986). Stream hydraulics as a major determinant of benthic invertebrate zonation patterns. *Freshwater Biology* **16,** 127–139.

Stewart-Oaten, A., W.W. Murdoch and K.R. Parker. 1986. Environmental impact assessment: "pseudoreplication" in time?. *Ecology* **67,** 929-40.

Strahler, A.N. 1964. Quantitative geomorphology of drainage basins and channel networks. Pages 39-76 in V.T. Chow (ed) Handbook of applied hydrology. 4th edition. McGraw-Hill, New York

Strahler, A.N. (1957). Quantitative analysis of watershed geomorphology. *Transactions of the American Geophysical Union* **38**, 91-920

ter Braak, C.J.F. and Verdonschot, P.F.M. (1995). Canonical correspondence analysis and related multivariate methods in aquatic ecology. *Aquatic Sciences* **57**, 255-289.

Terrell, C.R. and Perfetti, P.B. (1996). Water Quality Indicators Guide. Surface Waters. Terrene Institute. Washington DC.

Thienemann, A. (1925). Die Binnengewässer Mitteleuropas. Eine Limnologische Einführung. *Binnegewässer*. **1,** 1-125

Thoms M.C., Ogden R.W. & Reid M.A. (1999) Establishing the condition of lowland floodplain rivers: a palaeo-ecological approach. *Freshwater Biology*, **41**, 407-423.

Thoms, M.C. (1998). The condition of the Namoi river system. Cooperative Research Centre for Freshwater Ecology Technical Report, University of Canberra, Canberra, Australia.

Townsend, C.R., and A.G. Hildrew. (1994). Species traits in relation to a habitat templet for river systems. *Freshwater Biology* **31,** 265-275.

Tuffery, G. and Verneaux, J. (1968). Méthode de détermination de la qualité biologique des eaux courantes. Exploitation codifée des inentaires de la faune du fond. Ministère de l'Agriculture (France), Centre National d'Etudes Techniques et de Recherches Technologiques pour l'Agriculture, les Forêts et l'Equipment Rural, Section Pêche et Pisiculture 23p.

U.S. Environmental Protection Agency (U.S. EPA). 1990. *Biological criteria: National program guidance for surface waters.* (U.S. Environmental Protection Agency, Office of Water Regulations and Standards, Washington, D.C.) EPA 440-5-90-004.

U.S. Environmental Protection Agency (U.S. EPA). 1991a. *Biological criteria: State development and implementation efforts.* (U.S. Environmental Protection Agency, Office of Water, Washington, D.C.) EPA-440/5-91-003.

U.S. Environmental Protection Agency (U.S. EPA). 1991b. *Biological criteria: Guide to technical literature.* (U.S. Environmental Protection Agency, Office of Water, Washington, D.C.) EPA-440-5-91-004.

U.S. Environmental Protection Agency (U.S. EPA). 1991c. *Technical support document for water quality based toxics control.* (U.S. Environmental Protection Agency, Office of Water, Washington, D.C.) EPA 505-2-90-001.

U.S. Environmental Protection Agency (U.S. EPA). 1991d. *Biological Criteria: Research and Regulation: Proceedings of a Symposium.* (U.S. Environmental Protection Agency, Office of Water, Washington, D.C.) EPA-440-5-91-005.

U.S. Environmental Protection Agency (U.S. EPA). 1998. *Lake and reservoir bioassessment and biocriteria technical guidance document.* (U.S. Environmental Protection Agency, Office of Water, Washington, D.C.) EPA-841-B-98-007.

Underwood A. (1993). The mechanics of spatially replicated sampling programs to detect environmental impacts in a variable world. *Australian Journal of Ecology*. **18,** 99-116.

Underwood, A.J. (1994). On beyond BACI – sampling designs that might reliably detect environmental disturbances. *Ecological Applications*. **4,** 3-15.

Underwood, A.J. (1997). *Experiments in Ecology. Their Logical Design and Interpretation Using Analysis of Variance.* (Cambridge University Press, Cambridge.)

Van Sickle, J. and Hughes, R.M. (2000). Classification strengths of ecoregions, catchments, and geographic clusters for aquatic vertebrates in Oregon. *Journal of the North American Benthological Society.* **19**, 370-384.

Vannote, R.L., Minshall, G.W., Cummins, K.W., Sedell, R., and Cushing, C.E. (1980). The river continuum concept. Canadian Journal of Fisheries and Aquatic Sciences 37:130–137.Whittier, T.R., R.M. Hughes, and D.P. Larsen. (1988). Correspondence between ecoregions and spatial patterns in stream ecosystems in Oregon. Canadian Journal of Fisheries and Aquatic Sciences 45:1264-1278.

Verneaux, J., Galmiche, P., Janier, F., and Monnot, A. (1982). Une nouvelle méthode pratique d'evaluation de la qualité des eaux courantes. Un indice biologique de qualité général (BIG). Annales Scientifiques de l'Université de Franche-Comté Besançon, Biologie animal 4 ème série, fasc 3, Note No. 2:11-21.

Vinson, M. R. & C. P. Hawkins, (1996). Effects of sampling area and subsampling procedure on comparisons of taxa richness among streams. *Journal of the North American Benthological Society.* **15,** 392–399.

Washington, H.G. 1984. Diversity, biotic and similarity indices. A review with special relevance to aquatic ecosystems. Water Research 18: 653-94.

Wiederholm, T. (1980). Use of benthos in lake monitoring. *Journal of the Water Pollution Control Federation.* **52**, 283-297.

Woodiwiss, F.S. 1964. The biological system of stream classification used by the Trent River Board. *Chemistry and Industry* **11**, 443-7.

Wright J.F. (1995) Development and use of a system for predicting macroinvertebrates in flowing waters. *Australian Journal of Ecology*, **20**, 181–197..

Wright J.F., Moss D., Armitage P.D. & Furse M.T. (1984) A preliminary classification of running-water sites in Great Britain based on macroinvertebrate species and the prediction of community type using environmental data. *Freshwater Biology*, **14**, 221–256

Wright, J.F., Furse, M.T. and Moss, D. (1998) River classification using invertebrates: RIVPACS applications. *Aquatic Conservation: Marine and Freshwater Ecosystems.* **8**, 617-631.

Wright, J.F., Sutcliffe, D.W. and Furse, M.T. (2000). *Assessing the Biological Quality of Freshwaters: RIVPACS and Similar Techniques.* (Freshwater Biological Association, Ambleside.) 373 pp.

Wright, S. (1955). Limnological Survey of western Lake Erie. U.S. Fish. Wildl. Serv. Spec. Sci. Re. Fisheries No. 139.

Zarull, M.A. and Reynoldson, T.B. (1992) A management strategy for contaminated sediments: assessment and remediation. *Water Pollution Research Journal of Canada* **27** 871-882.

Index

MIX
Papier aus verantwortungsvollen Quellen
Paper from responsible sources
FSC® C105338

If you have any concerns about our products,
you can contact us on
ProductSafety@springernature.com

In case Publisher is established outside the EU,
the EU authorized representative is:
Springer Nature Customer Service Center GmbH
Europaplatz 3, 69115 Heidelberg, Germany

Printed by Libri Plureos GmbH
in Hamburg, Germany